Free
Calculus

$$\left(\begin{array}{c}\text{A Liberation} \\ \text{from Concepts} \\ \text{and Proofs}\end{array}\right)$$

Free
Calculus

(A Liberation from Concepts and Proofs)

Qun Lin
Chinese Academy of Sciences, China

 **World Scientific**

NEW JERSEY · LONDON · SINGAPORE · BEIJING · SHANGHAI · HONG KONG · TAIPEI · CHENNAI

Published by

World Scientific Publishing Co. Pte. Ltd.

5 Toh Tuck Link, Singapore 596224

USA office: 27 Warren Street, Suite 401-402, Hackensack, NJ 07601

UK office: 57 Shelton Street, Covent Garden, London WC2H 9HE

British Library Cataloguing-in-Publication Data
A catalogue record for this book is available from the British Library.

FREE CALCULUS
A Liberation from Concepts and Proofs

ISBN-13 978-981-270-458-0
ISBN-10 981-270-458-2

Typeset by Stallion Press
Email: enquiries@stallionpress.com

Printed in Singapore by Mainland Press Pte Ltd

Free Calculus—A Liberation from Concepts and Proofs

Qun Lin

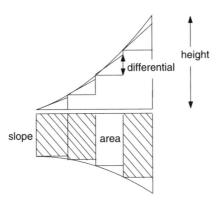

Target audience

1. Undergraduate; 2. Preparatory student; 3. High school student; 4. The public.

Chapter 0 is an introduction of Free Calculus for beginners (in particularly for students of humanities, management and business majors, and of high school level) with plain English and without mathematical symbols. Chapter 1 is a mathematical representation of Free Calculus, which is written and enough for general undergraduates (as well as high school level students) and is a first period course for engineering and science majors. Chapters 2 and 3 are about the differential equations and are the second period course for engineering and science majors. Chapter 4 is a possible generalization of calculus, written for students who still have vigor and interest to learn more calculus.

Characteristic

Free Calculus has a difference from other texts. It uses plain English, instead of mathematical symbols, to give a definition property of the differential and the rigorous (instead of intuitive) proofs of calculus theorems by a few lines of arithmetic (or calculation) through the definition itself, without more concept and tricky techniques. So it is a liberation from concepts and proofs. This makes the physics teaching in step with the calculus teaching.

Preface

Standard calculus texts involve a substantial amount of *a priori* knowledge and intricate technique, requiring more than 100 pages for a complete proof. Students find it too difficult and too long to learn, and easily forgotten. An idealist would asks: "Can we use plan English (instead of mathematical symbols) and, at the same time, give rigorous (instead of intuitive) proofs of calculus theorems with just a few lines of arithmetic (or calculation), without *a priori* knowledge and tricky techniques? An "Unarmed" calculus does not exist but it only needs a definition property of the "differential." If this definition can be explained well, we would have arrived at the proofs for the fundamental theorem (FT) and Taylor theorem (TT).

$$FT = a\ definition + few\ lines\ of\ arithmetic$$
$$TT = few\ lines\ of\ calculation\ from\ FT$$

In the first part: Plain English Calculus, we concentrate on the explanation of the "differential", in an as elementary manner as possible. With just a few lines of proof, this will lead to calculus theorems.

Subsequently, the second part translate plain English into the function language to tally with the formal definition and theorems in standard texts. All together, the fundamental essence, including rigorous proofs of the FT and TT is done in less than 10 pages.

This is, of course, the statement of only one of the parties. In fact, almost all of the calculus experts frown upon this approach. Only few colleagues express admiration and use this book for students majoring in physical culture and liberal arts. Nevertheless, it is still useful as an introductory course for the physics and engineering cohort.

Part I: Plain English
(with rigorous definition and proof)

Calculus' source and definition property of tangent line

One day I walked under an old tree. A tourist guide told sightseers that this tree grows up every year and a surveyor measures its height every year. I knew from middle school trig that the height can be measured in terms of one slope through a tangent formula:

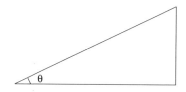

without cutting the tree! A breakthrough is to change the tree to a hill where one faces a curved hypotenuse instead of the straight one:

From a special (straight) to the general (curve).

But, in calculus, the curved hypotenuse is divided into many short segments, each of which is replaced by a tangent line (a straight line closest to the curve near the starting node, if exists) with a height named by

$$\text{differential} = (\text{starting slop}) (\text{base}).$$

Each differential, however, contains a measured error to the true height:

$$\text{True height} = \text{differential height} + \text{measured error}$$

(see Fig. 1). A problem arises: How small the measured error is? Can we guarantee it as small as possible? The measured error is nothing but a distance from the tangent (if exists) to the curve, much smaller than that from other secants (by the property of the tangent: A straight line closest to the curve near a node).

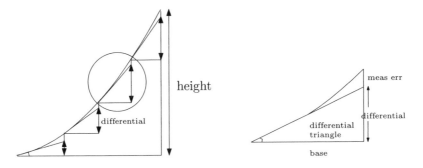

Figure 1. From the general to a special.

Since the latter (the distance from the secant to the curve) is almost the same as the small base:

$$\frac{\text{secant measured error}}{\text{base}} \approx \text{const}$$

(see the explanation at the end of this section), the former (the distance from the tangent to the curve) must be much smaller than the small base: Correctly speaking, the ratio of the tangent measured error to the small base,

$$\text{relative error} := \frac{\text{measured error}}{\text{base}}$$

is also small: as small as you want if reduce the base (denoted $\ll 1$ for simplicity),

i.e., the measured error of the tangent is not proportional to the small base but reducing faster than the base. This requirement is regard as the definition property

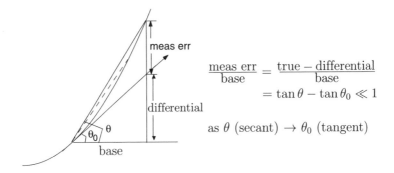

$$\frac{\text{meas err}}{\text{base}} = \frac{\text{true} - \text{differential}}{\text{base}}$$
$$= \tan\theta - \tan\theta_0 \ll 1$$

as θ (secant) $\rightarrow \theta_0$ (tangent)

Figure 2. Tangent (θ_0) is the limit position of secants (θ).

of the tangent line (in the pointwise sense). In fact, for any secant with a slope $\tan\theta$ (with $\theta \neq \theta_0$):

$$\text{Secant relative error} = \frac{|(\tan\theta)(\text{base}) - \text{true}|}{\text{base}} \rightarrow |\tan\theta - \tan\theta_0| = \text{const} \neq 0.$$

Comparing with

$$\text{tangent relative error} \ll 1$$

we distinguish between the tangent and the secant.

To sum up, if the tangent exists,

$$\frac{\dfrac{\text{Secant measured error}}{\text{base}} \approx \text{const}}{\dfrac{\text{Tangent measured error}}{\text{base}} \ll 1.}$$

This is the uniquely fundamental concept of calculus in our book and the student must understand it thoroughly.

A rigorous enough proof of FT

From Fig. 1

$$\text{total true height} = \text{sum of differentials} + \text{total error}$$
$$\text{total error} = \text{sum of measured errors}$$

(over all segments). With a few lines of arithmetic we get an identity:

$$\text{total error} = \text{sum of (relative errors) (bases)}$$
$$= \text{sum of} \left(\frac{\text{bases}}{\text{sum of bases}} \text{ relative errors}\right)(\text{sum of bases})$$
$$= (\text{average of relative errors}) (\text{sum of bases})$$

(over all segments), proportional to the average:

$$\text{Total error} \approx \text{average of relative errors.}$$

If and only if the latter is small or, simply, the relative errors are uniformly small (this can be regarded simply as a definition property of the tangent line in the global sense), the total error $\ll 1$: as small as you want that can be eliminated, then we get an ideal equality:

Total true height = integral of differentials.

This is the FT of calculus. Thus the differential (the height of the differential triangle in Fig. 3) is an element of the FT (the total height of curved triangle), whereas the FT is a superposition (or integral) of differentials.

To sum up, if and only if the average is small or if the relative errors are uniformly small, we have the FT.

This completes a rigorous enough proof of the FT of a few lines arithmetic, with plain English and without mathematical symbols and tricky techniques.

Figure 1 is the single most natural figure to discover calculus with cartoon watching instead of words, and is often used for calculus popularization in China since the 1990s (Lin, 1997).

Second Part: Function Language

(with formal calculus)

In order to identify the above statement with standard texts we now have nothing to do, just a translation from plain English to the function language.

Derivative definition

The concept of the tangent slope or differential is determined by the derivative definition. Let $f = f(x)$ be a given function (producing a curve) defined on the interval $[a, b]$. Then, over each subinterval $[x, x + h]$, for all points $x + h$ near the node x, we have

True height: $f(x + h) - f(x)$

Tangent slope $\tan \theta_0$ at node x: Derivative $f'(x)$

Differential: $f'(x)h$

Measured error: $f(x + h) - f(x) - f'(x)h$

Relative error $\epsilon(x, h) := \frac{f(x+h)-f(x)}{h} - \frac{f'(x)h}{h}$

Each measured error = (relative error) (base):

$$f(x + h) - f(x) - f'(x)h = \epsilon(x, h)h$$

If over each subinterval $[x, x+h]$, $|\epsilon(x,h)| \ll 1$ or simply

$$\epsilon(x,h) \ll 1$$

the derivative f' is uniquely defined pointwisely. The notation $\epsilon(x,h)$ will be explicit for concrete functions or the three test functions $f = x^n$, $\sin x$, e^x (hence their arithmetic and composites, and all elementary functions):

$$\epsilon(x,h) \le Ch$$

(named by the Lipchitz assumption).

One line rigorous proof of FT

The standard proof of the FT uses a lot of *a priori* knowledge (the inf-sup theorem in the real number theory, the extreme value theorem of continuous functions, the mean value theorem, ...) and tricky techniques. Students learn it and then forget it. Instead, we use (Lin, 1977–2005) only the differential definition itself and its superposition to obtain the FT without more *a priori* knowledge and tricky techniques. In fact, based on the height measurement or the differential version on each subinterval $[x, x+h]$, one can calculate the total true height of f on $[a,b]$ through a superposition involving h^{-1} terms and get the total error:

$$f(b) - f(a) - \text{sum of } f'(x)h = \text{sum of } \epsilon(x,h)h$$
$$= (b-a)\left[\frac{h}{b-a}\text{sum of } \epsilon(x,h)\right],$$

proportional to the average. If and only if the average is small or if the relative errors are uniformly small, or simply the upper relative error

$$\epsilon(h) = \text{upper } \epsilon(x,h) \ll 1$$

(which, combining with the pointwise definition of the derivative in the last section, can be regarded simply as the global definition of the derivative).

Few lines direct proof of TT (as a corollary of FT)

The standard proof of the TT uses indirect "integration by parts", spending much thinking and time. Students, however, prefer a direct (mechanical) calculation through the FT itself as follows:

Constant approximation

$$f(x+h) - f(x) = \int_x^{x+h} f'(s)\,ds :\approx h \underset{[x,x+h]}{\text{upper }} |f'|,$$

Linear approximation

$$f(x+h) - f(x) - hf'(x) = \int_x^{x+h} [f'(s_2) - f'(x)]ds_2$$

$$= \int_x^{x+h} \int_x^{s_2} f'(s_1)\,ds_1 ds_2$$

$$:\approx \underset{[x,x+h]}{\text{upper }} |f'| \int_x^{x+h} \int_x^{s_2} ds_1 ds_2$$

$$= \frac{h^2}{2} \underset{[x,x+h]}{\text{upper }} |f''|,$$

Quadratic approximation

$$f(x+h) - f(x) - hf'(x) - \frac{h^2}{2} f''(x)$$

$$= \int_x^{x+h} \int_x^{s_3} [f''(s_2) - f''(x)]ds_2 ds_3$$

$$= \int_x^{x+h} \int_x^{s_3} \int_x^{s_2} f'''(s_1)\,ds_1 ds_2 ds_3$$

$$:\approx \underset{[x,x+h]}{\text{upper }} |f'''| \int_x^{x+h} \int_x^{s_3} \int_x^{s_2} ds_1 ds_2 ds_3$$

$$= \underset{[x,x+h]}{\text{upper }} |f'''| \frac{1}{3!} \left(\int_x^{x+h} ds \right)^3 = \frac{h^3}{3!} = \underset{[x,x+h]}{\text{upper }} |f'''|,$$

and so on, where we use the fact that if $|g| \leq C$ (constant) then $|\int g(s)ds| \leq C \int ds$ (easily verifiable). This completes the direct proof of the FT−I believe it is the simplest.

History

Such an approach, using the regular pointwise definition of the derivative, accompanying with the uniform assumption (or even the concrete Lipschitz assumption), to give rigorous proofs of the FT and TT of calculus (including abstract calculus) with a few lines of calculations, has appeared previously in my early booklets (since 1990's). When I'm writing this new one, a friend, Michael Livshits, told me that the uniform Lipschitz assumption, have been used by himself (2004) and Hermann

Karcher (2002). Then I found in the internet another book ≪ Applied Calculus ≫ (by Karl Heinz Dovermann, July 1999) with a story about the pointwise Lipschitz assumption: "This approach, which should be to easy to follow for anyone with a background in analysis, has been used previously in teaching calculus. The author learned about it when he was teaching assistant for a course taught by Dr. Bernd Schmidt in Bonn about 20 years ago. There this approach was taken for the same reason, to find a less technical and efficient approach to the derivative. Dr. Schmidt followed suggestions which were promoted and carried out by Professor H. Karcher as innovations for a reformed high school as well as undergraduate curriculum. Professor Karcher had learned calculus this way from his teacher, Heinz Schwarze. There are German language college level textbooks by Kutting and Moller and a high school level book by Muller which use this approach." On the other hand, the general uniform definition of the derivative has have been used since 1940's in many books (e.g., Vainberg (1946,1956), Ljusternik–Sobolev (1965), Lax, etc. (1976), Lin (1999, 2002), Karcher (2002), Livshits (2004), Zhang, 2006).

Special Acknowledgements

The author would like to express his thanks to Prof. K. K. Phua whose support and help play a decisive part in writing and publishing this book.

Qun Lin

Beijing, November 2006

Contents

Chapter 0

Calculus in Terms of Images:

Hill-Climbing

This is an introduction of calculus for beginners and a review for students who have already known calculus. Here calculus is motivated by two measurements, slope and height, when climbing a hill. In other words, calculus is regarded as a curved trig beginning with the curved triangle (see Refs. 15–20), measuring the height with slopes.

A student (A) has a dialogue with a teacher (B) about differential calculus and integral calculus when they are climbing a smooth hill.

0.1. Hill Behavior and Slope

They climb up, and rest on a platform when they are tired. Then they climb up again.

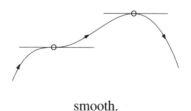

smooth.

A: Is this platform a stationary point in differential calculus?

1

B: It is! Since the hillside is rising before the platform and rising again after the platform, this is called a stationary point.

They arrive at the second platform and climb down then.

A: Is such a platform a local maximum in differential calculus?

B: It is! Since the hillside is rising before the platform but falling then, this is a top and called a local maximum.

A: So far we touch only some feels about rising, falling, platform or local maximum and local minimum, without mathematics.

B: Mathematics, or differential calculus, is invented to measure the feeling of climbing quantitatively (Bruter, 1973), with a parameter called the tangent slope (which was used in Section 1 of the Preface), corresponding to the sign > 0 (rising), < 0 (falling) or $= 0$ (platform), the gradient (steepness or gentleness) of rising or falling, and the turning point of bend.

A: The slope parameter is so fundamental for climbing.

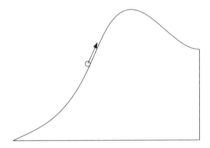

B: The slopes describe how curve a function is. Generally speaking, the slope is invented to describe the changing world, e.g., economy states, raising or falling, fast or slow, or stable. For instance, Arnold uses the signs of slopes to observe the states of economy.[2]

0.2. Hill Height and Slope: Unconstructive Tangent Formula

A: The slope is so fundamental to measure the behavior of a hill. Can the slope be also used to measure the height of a hill?

B: In the Preface, the tangent slope was used to measure the differential (=(starting slope) (base)). But how does one measure the horizontal base? It is almost impossible! So, the FT method is not practical. Indeed, nobody uses the FT to measure the hill height. The FT is to establish the ties between the height and slopes, just like Pythagoras theory establish the ties between the hypotenuse and the other two edges.

A: Do we have other ways to measure hill height?

B: It has been seen in the regular trig that the slope can be used to measure the height of a tree:

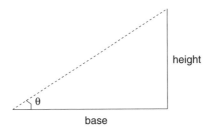

If a hill has a straight hillside (or hypotenuse), this is just a regular trig measurement (tangent formula).

A: But hills usually have a curved hillside (or curved hypotenuse).

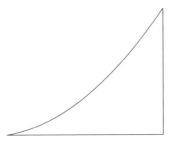

Can we still measure the height (unknown) of a hill a single slope and a single base?

B: But the slope of the hillside is changeable; different points would have different slopes: Which one do we assign?

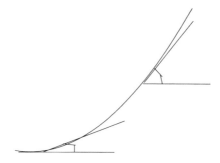

If we randomly select a point, e.g., again the starting point, through the slope at which to compute the height, a large error will occur.

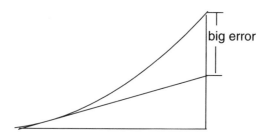

A: Does there indeed exist such a slope to give the hill height?

B: Since the hill exists, with the foot and top

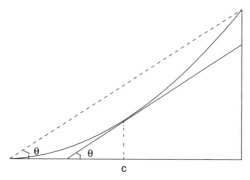

Curved and straight (secant) right triangles have the same height,
base, and slope (θ).

there exist a secant line linking the foot and top and, intuitively, a particular tangent line on the smooth hillside parallel to this secant line, with the same slope. Hence, we get the true

$$\text{height} = (\text{secant slope})(\text{base}) = (\text{tangent slope at } c)(\text{base}),$$

a tangent formula for the curved right triangle, or called the Mean Value Theorem (MVT) in differential calculus. Thus if we want to link the height with a slope, we should use the MVT. It is geometrically evident, but a complete proof of this theorem is very long and is best left to advanced calculus.

A: Does the mean value c is knowable?

B: No. We are near-sighted. We don't see far away. We don't know where the value c is. Such a tangent formula is existing somewhere but we do not know where it is. It is not an algorithm. So the MVTs is an unconstructive tengent formular.

A: It is a pity that such a beautiful theorem is useless in practical sense!

B: Indeed, my friend Michael Livships hates the MVT, which controls a long and an unknown mean value!

0.3. Review for FT

B: Let us recall the inventing process of the FT, a typical Cartesian methodology. The first step is to shorten the whole unit into one segment. Let the curved hypotenuse be shortened into one segment.

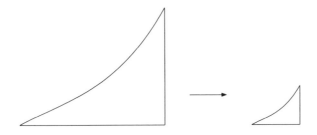

The curved triangle is then shortened into a small one.

A: It seems that we have won nothing from the first step. Curved is still curved.

B: However, within the shortened curve segment the slopes at all points are nearly the same. Now choose an arbitrary point within the curve segment, e.g., the starting point; the slope at this point (called the starting slope) is the slope of the curve segment,

Curve (or curved triangle) $\approx$ tangent (or differential triangle).

where within the curve segment

$$\text{height variation} \approx (\text{starting slope})(\text{base})$$

called a differential (measurable).

A: The quantity, differential, thus obtained is the height of the tangent line within the curved segment (or the height of the differential triangle), rather than the true variation (unmeasurable).

B: But the error should be small, even when compared with the small base.

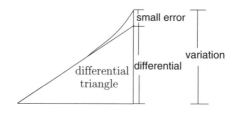

$$\text{variation} \approx \text{height of differential triangle}$$
$$(\text{by tangent formula}) = (\text{slope})(\text{base})$$
$$(\text{named by}) = \text{differential}.$$

More accurately, the ratio of this error to the small base, the relative error, is still small. This is a key step (the second step), where by doing differential we have nearly gotten the height variation within one segment.

A: In the above step, within one segment the curve is replaced by the tangent line rather than the secant. Why?

B: Standard textbooks used the secant to replace the curve (within a short segment) but then used the MVT to replace the secant slope with the tangent slope at an unknown mean value, appending unnecessary twists; so, why not use directly the tangent slope at the starting point (agreeing with Euler's constructive algorithm in Chap. 2) without the replacement. The latter produces an error (very small) but avoids the MVT (with a long proof and an unknown mean value). This is indeed a definition of the differential.

A: So far we have only solved the height variation within the initial segment, or just have completed a differential computation (or differentiation). How to extend the initial segment to the whole unit to complete the integral computation (or integration)?

B: The initial segment, together with the differential triangle, is extended segment by segment.

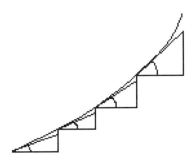

Add up the differential heights that we have obtained through all differential triangles within every segment (where we use all the slopes) and the resulting value will serve as the total true height of the curved triangle.

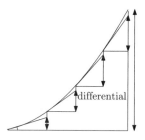

Total true height $\approx$ sum of differentials.

A: This value, however, is not the true height itself, the error consists of the accumulation of differential errors within every segment. Can we expect this value would still be near the true height of the hill, and the finer the better?

B: This is a bold and adventuristic expectation because the accumulation of differential errors, the total error, still may not be small, unless every differential error is so small (e.g., n^{-2} when base $= n^{-1}$) that even their accumulation (involving n terms) is still small (of n^{-1}). This can be achieved by the definition property of the tangent line mentioned in Section 1 of Preface.

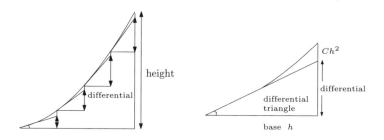

A: When does close ($\approx$) become exact ($=$)?

B: Now it is possible to eliminate the total error by the infinitesimal method, where the total curved triangle is replaced by infinitely many infinitesimal differential triangles and the total true height is replaced by infinitely many infinitesimal differentials (and so all the slopes):

$$\text{total true height} = \text{sum of (differentials} + O(n^{-2}))$$
$$= \text{sum of differentials} + O(n^{-1});$$

hence close ($\approx$) becomes exact ($=$) for small base.

total true height $=$ integral of differentials

each differential $=$ (starting slope)(base).

This is indeed a quick proof of the FT. More refined proof will require the use of the average relative error, see Section 2 of Preface. It is fundamentally constructive by differentials. It is not only a computed method for the curve height but gives the relationship between different quantities, e.g., between area and height. See Sec. 0.5.

A: I understand now. The FT is nothing but differentiation first and integration afterwards: shorten the curve, it becomes a differential or a tangent formula; draw the curve, it becomes an integral or the FT. In short,

shortening $=$ tangent formula; drawing $=$ FT.

Essentially, integral calculus (university) is nothing more than a tangent formula (high school). When drawing, tangent formula (high school) becomes the FT (university).

B: To sum up, when you climb a hill you do a differential in each step; when you arrive at the top you complete an integral.

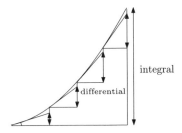

The FT has already been fished out.

0.4. Hillside Length and Slope: Pythagoras Theorem

A: How do we measure the arclength? Can the length of the hillside be also measured by slopes?

B: If the hillside is straight, this is nothing but the Pythagoras formula,

$$\text{length} = \sqrt{1 + (\text{slope})^2} \ \text{base}.$$

Within one segment, the length of the curved hillside (called arclength) can be approximately measured using the starting slope.

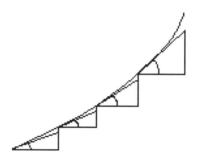

$$\text{arcdifferential} = \sqrt{1 + (\text{starting slope})^2} \ \text{base}$$

(i.e., Pythagoras' formula holds on a short segment). Adding up gives

whole arclength = integral of infinitely many infinitesimal arcdifferentials.

Arnold gives a remarkable example showing that the length of a sine curve increases by 20% only than that of the cross axle, e.g., a 5 m axle corresponds to a 6 m sine curve.[2]

Thanks to the arclength formula which tells us quantificationally how long the curved road is! This is a real problem which needs the slop's help.

0.5. Area and Slope

A: The slope parameter prevails everywhere in trigonometry and calculus. Can we use the slope to measure areas enclosed by some curves?

B: Since the slopes are the fundamental quantity of a curve let us record them and get a slope curve, the inverted image of the original (height) curve. It is amazing that the area enclosed by the slope curve is equal to the height of the original curve, as seen in the following height–slope figure:

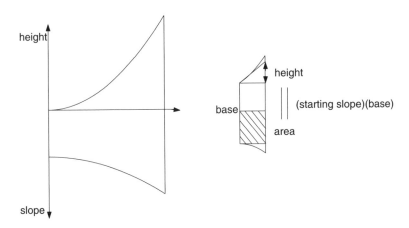

So, such an area can be computed by the height of the original curve:

area enclosed by the slope curve = height of the original curve.

A: This figure also appeared in a paper from the United States in 2003 with comments: "it is probably the single most important figure in this book. It's a picture worth a thousand symbols and equations, encapsulating the essence of integration in a single snapshot ... If you understand only half of what I've just written, you're way ahead of most students of calculus."

0.6. Explaining All of Calculus in a Single Figure

B: It was my goal to explain all of calculus in a single figure. Such a single figure first appeared in Lin's Calculus Cartoon, Chinese Dailies, 1997 (see also Lin's English print),[15] and appeared in the cover of Lin's *Calculus Cartoon*, 1999,[16] and then reappeared in the text, *Calculus for Dummies*, 2003.[23]

A: Besides solving the problems from the geometry measurement, can calculus be used for other fields? Literature students would think they read novels without knowing calculus.

0.7. Calculus and Novels

B: The calculus spirit, differentiation first and integration afterwards, seeps into all fields, even into Tolstoy's "War and Peace."[26] In fact, Tolstoy said in his book: "only by taking an infinitesimally small unit for observation (the differential of history, that is, the individual tendencies of men) and attaining to the art of integrating them (that is, finding the sum of these infinitesimals) can we hope to arrive at the laws of history." How do you understand what he said if you do not understand the calculus spirit? You will know that the story was so but not why it was so.

A: Are there more examples in humanities?

B: A car engineer proposed a plan to vitalize car industry in China, named "differentiation first and integration afterwards." This is a general principle in the management science. That is to divide an unmeasurable system into many measurable units (differentials), and that small errors that arise in subsystems will not cause any large error output of the whole system. Or, details (differentials) are the most important, if you do well for details, so do for the whole system.

A: Such arguments are far-fetched.

B: The calculus language is more credible for natural sciences, in which a derivative is a rate (velocity), and perhaps the rate (velocity) concept is easier to be understood than the slope for a beginner. But for natural sciences we need differential equations to express their laws and to solve them. See Chap. 3.

A: What is a differential equation? We only know algebra equations in high schools.

B: The simplest differential equation is the FT, knowing slopes and solving for the height. The differential equation approach is a revolution in natural sciences, where Newton used it first and then all scientists (including Maxwell and Einstein) follow him.

A: So far, calculus is dramatic, developing acts by acts. What is its future?

B: A big subject is about multivariable differential equations and their numerical methods. Our dialog is a drop in the mathematical bucket.

A: I cannot accept so many materials. We must stop now.

B: Thank you, my daughter, for your role A and your enquisition from start to finish.

Chapter 1

Official Calculus:

Differentiation and Integration

> *Outline: Chapter 1 = two inequalities = differential inequality +*
> *fundamental inequality. The latter is the sum of the former (over*
> *all subintervals). And hence, finally, Chap. 1 = one inequality.*

Calculus in this book only admits a definition of the differential [or derivative, see (1.1)] without other prior knowledge. The FT, in one word, is the sum of these definitions without more proofs. This has been described in using plain English in the first part of the Preface. This chapter will use the function language in order to get more conclusions (e.g., the TT) and applications.

1.0. A Case: Height and Slopes

Trigonometry begins with measuring the height (unknown) of a tree using a slope (what we know) or the direction of a dummy hypotenuse. This problem leads to a tangent formula.

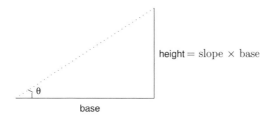

$$\text{height} = \text{slope} \times \text{base}$$

A regular triangle.

Parallely, calculus begins with measuring the height of a hill using the slopes of the curve hillside, as a curved trig (see Refs. 15–20). This leads to the Newton–Leibniz formula.

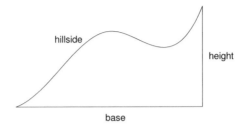

A curved triangle.

But calculus need more concepts e.g. the tangent slope, differential, measured error, relative error and total error, as indicated in the Preface with the following figures:

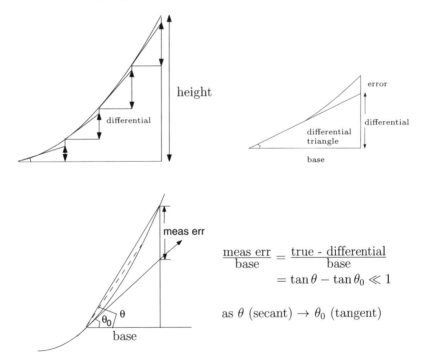

$$\frac{\text{meas err}}{\text{base}} = \frac{\text{true - differential}}{\text{base}}$$

$$= \tan\theta - \tan\theta_0 \ll 1$$

as θ (secant) $\to \theta_0$ (tangent)

we translate them into the function languages as follows.

1.1. Translating into Function Language

Let $f = f(x)$ be a given function (producing a curve) defined on the interval $[a, b]$. Then, over each subinterval $[x, x + h]$, for all points near the node x.

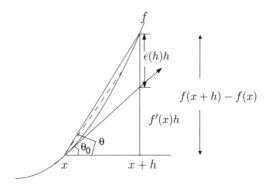

True height: $f(x + h) - f(x)$

Tangent slope $\tan \theta_0$ at node x: Derivative $f'(x)$

Differential: $f'(x)h$

Each measured error = (relative error)(base):

$$f(x + h) - f(x) - f'(x)h = \epsilon(x, h)h$$

If over each subinterv $[x, x + h]$

$$\epsilon(x, h) \ll 1 \qquad (1.1)$$

the derivative f is uniquely define pointwisely.

If the upper relative error $\epsilon(h) \ll 1$ is independent of the node x, a continuous derivative f' is defined, and (1.1) is called as the first inequality. From this point onward, we will frequently use the uniform bound, $\epsilon(h)$, to replace the pointwise bound, $\epsilon(x, h)$.

Total error $\approx$ upper relative error:

$$f(b) - f(a) - \text{sum of } f'(x)h = \text{sum of } \epsilon(x, h)h$$
$$= (b - a)\frac{h}{b - a} \text{ sum of } \epsilon(x, h)$$
$$\leq (b - a)\epsilon(h), \tag{1.2}$$

$$\epsilon(h) = \text{upper } \epsilon(x, h),$$

called the second inequality. If and only if the average is small, or simply the upper is small, see (1.1), then so is the total error. From which an ideal equality follows:

$$f(b) - f(a) = \int_a^b f'(x)dx. \tag{1.3}$$

Derivative's construction. The above versions for the derivative definition can be used to construct the derivative in an expansion way: Expand the true height, $f(x + h) - f(x)$, into a term of the first power of h and a remainder that is reducing faster than h, then the coefficient of the first power term is chosen as f' (uniquely). E.g. $f(x) = x^2$, its true height is $(x+h)^2 - x^2 = 2xh + h^2$, then the coefficient of the first power of h is chosen as $f'(x) = 2x$ (uniquely) and the remainder $= h^2$ is reducing faster than h, or the relative error $\epsilon(h) = h$. Similarly, for concrete functions (x^n, $\sin x$, e^x (hence any elementary function), we can construct this way the concrete f' and $\epsilon(h)$ (uniquely) where the relative error $\epsilon(h)$ is inherently independent of the node x and is a multiple of h:

$$\epsilon(h) \leq Ch, \tag{1.4}$$

where an abstract $\epsilon(h)$ is replaced by a concrete Ch, simpler and better, and so for all measured errors (1.1) and the total error (1.2):

$$|f(x + h) - f(x) - f'(x)h| \leq Ch^2$$

$$|f(b) - f(a) - \sum f'(x)h| \leq Ch.$$

They are also better for the beginner to understand before learning limits. However, for proving the FT, (1.3), the minimal assumption

$$\epsilon(h) \to 0 \quad \text{as } h \to 0 \tag{1.5}$$

is enough while the strengthened assumption (1.4) is more than enough. So we use (1.5) when proving the FT and use (1.4) when the beginner does not understand the limits.

Let us supplement more examples for the derivative's construction.

(i) When $f(x) = x^n$ $(n = 0, 1, 2, \ldots)$, its true height

$$(x + h)^n - x^n = nx^{n-1}h + \frac{n(n+1)}{2!}x^{n-2}h^2 + \cdots + h^n$$

then choose the coefficient of the first power term, $f'(x) = nx^{n-1}$ (uniquely), to make the measured error

$$\left| \frac{n(n+1)}{2!}x^{n-2}h^2 + \cdots + h^n \right| \leq Ch^2$$

reducing faster than h, or the relative error $\epsilon(h) \leq Ch$.

(ii) When $f(x) = \sin x$ then choose $f'(x) = \cos x$ (uniquely) to make the measured error reducing faster than h: Assume that $h > 0$ (then you can check by yourself when $h < 0$),

$$
\begin{aligned}
|\sin(x + h) - \sin x - f'(x)h| &= |\sin x \cos h + \cos x \sin h \\
&\quad - \sin x - \cos x \cdot h| \\
&= |\sin x(\cos h - 1) + \cos x(\sin h - h)| \\
&\leq 1 - \cos h + (h - \sin h) \leq Ch^2.
\end{aligned}
$$

Indeed, this has been examined in standard texts (e.g. Strang's Calculus, p.65)[25]:

$$1 - \cos h = 2 \sin^2 \frac{h}{2} \leq \frac{1}{2}h^2, \quad h - \sin h < h - h \cos h \leq \frac{1}{2}h^3$$

or the relative error $\epsilon(h) = \frac{1 - \cos h}{h} + \frac{h - \sin h}{h} \leq Ch$.

Similarly, when $f(x) = \cos x$ then choose $f'(x) = -\sin x$ (uniquely) to make the measured error $\leq Ch^2$ reducing faster than h, or the relative error $\epsilon(h) \leq Ch$;

For the following functions, e^x and $\ln x$, whose definitions use limit notions, we must admit $e = \lim_{n \to \infty}(1 + \frac{1}{n})^n$. (It is best left to next volume.)

(iii) When $f(x) = e^x$ then choose $f'(x) = e^x$ (uniquely) to make the measured error reducing faster than h: Assume that $h > 0$ (then you can check by yourself when $h < 0$),

$$e^{x+h} - e^x - e^x h = e^x(e^h - 1 - h) \leq e^b(e^h - 1 - h) \leq Ch^2.$$

Indeed, this has been examined in standard texts (e.g. Strang's Calculus, p.233, 255):

Since $e = \lim_{n \to \infty} (l + \frac{h}{n})^{\frac{n}{h}}$,

$$e^h = \lim_{n \to \infty} \left(1 + \frac{h}{n}\right)^n = \lim_{n \to \infty} \left(1 + h + C_n^2\left(\frac{h}{n}\right)^2 + \cdots + C_n^n\left(\frac{h}{n}\right)^n\right),$$

$$e^h - 1 - h = \lim_{n \to \infty} \left(C_n^2\left(\frac{h}{n}\right)^2 + \cdots + C_n^n\left(\frac{h}{n}\right)^n\right)$$
$$= h^2 \lim_{n \to \infty} \left(C_n^2\left(\frac{1}{n}\right)^2 + \cdots + C_n^n\left(\frac{1}{n}\right)^n h^{n-2}\right)$$
$$\leq h^2 \lim_{n \to \infty} \left(C_n^2\left(\frac{1}{n}\right)^2 + \cdots + C_n^n\left(\frac{1}{n}\right)^n\right)$$
$$= h^2(e - 2)$$

or the relative error $\epsilon(h) = e^b\left(\frac{e^h - 1}{h} - 1\right) \leq Ch$.

Similarly, when $f(x) = \ln x$ then choose $f'(x) = \frac{1}{x}$ (uniquely) to make the measured error reducing faster than h, or the relative error $\epsilon(h) \leq Ch$.

We can, in principle, examine more concrete functions. It will be too expansive, however, if we need to find the derivative f' and the relative error $\epsilon(h)$ for each concrete function. Calculus is so successful because it relies heavily on a very few functions, e.g. the polynomial, the sine and cosine, the ln and exponential functions, even general power functions (see subsequent Sec. 1.4). Instead of examining more functions, we only need to examine these few functions.

Those concrete functions approve the definition, or the first inequality (1.1): For all points $x + h$ near x

$$|f(x + h) - f(x) - f'(x)h| \leq \epsilon(h)h,$$

i.e., the remainder, or the upper relative error $\epsilon(h)$, is inherently independent of the node x.

Direct proof of general FT Through the above first inequality (1.1), adding up, we get the second inequality, (1.2), controlling the total error, or even the generalized second inequality

$$\left| f(b) - f(a) - \sum_{\xi \in [x, x+h]} f'(\xi)h \right| \leq (b - a)\epsilon^*(h),$$

where $\epsilon^*(h) \approx \epsilon(h)$. If and only if the average of $\bar{\epsilon}$ in (1.8) is small or if the upper relative error $\epsilon(h)$ is chosen small then the total error is still small, and smaller the upper relative error $\epsilon(h)$, closer to the same value, $f(b) - f(a)$ (when the sums $\sum_{\xi \in [x, x+h]} f'(\xi)h$ are calculated successively) and finally, in the ideal or limit case, close ($\approx$) becomes exact ($=$):

$$f(b) - f(a) = \int_a^b f'(x)dx$$

where we use the notation, integral over $[a, b]$, to denote a sequence of the sums which closes to the same value, $f(b) - f(a)$, no matter how fine the subdivision and where the intermediate point ξ. This is the FT of calculus, derived from a superposition (or nested sum) of the definition (1.1) itself with few lines calculation and without more *a priori* knowledge.

The FT (1.3) is a powerful tool to produce everything in calculus, e.g., if $f' \equiv 0$, then $f \equiv c$; if $f' \geq 0$, then $f \uparrow$. More important: In reverse order, the consequence (1.3) can be used to rewrite the causality (1.1) and extend it to Taylor's theorem (TT) as follows.

Rewriting implicit (1.1) into explicit. Use (1.3) itself to rewrite (1.1) in an integral equality:

$$f(x + h) - f(x) = \int_x^{x+h} g(s)ds,$$

and (1.1) is rewritten as

$$f(x + h) - f(x) - hf'(x) = \int_x^{x+h} [f'(s) - f'(x)]ds$$

with the explicit upper relative error

$$\epsilon(h) = \underset{[x,x+h]}{\text{upper}} |f'(s) - f'(x)|,$$

where we use the fact that if $|g| \leq C$(constant) then $|\int g(s)ds| \leq C \int ds$ (with $g = f'$). Why? From the definition (1.1)

$$\int_x^{x+h} g(s)ds = g(x)h + \epsilon(x, h)h, \quad |\epsilon(x, h)| \leq \epsilon(h),$$

$$\int_a^b g(s)ds = \sum \int_x^{x+h} g(s)ds = \sum g(x)h + \sum \epsilon(x, h)h,$$

$$\left| \int_a^b g(s)ds \right| \leq \left| \sum g(x)h \right| + \left| \sum \epsilon(x, h)h \right|$$
$$\leq \sum |g(x)| h + \sum |\epsilon(x, h)| h$$
$$\leq C(b - a) + \epsilon(h)(b - a),$$

$$\left| \int_a^b g(s)ds \right| \leq C(b - a).$$

Direct proof of TT. The FT (1.3) itself (instead of indirect "integration by parts"), when f is smooth enough, is successively used to represent the remainder of TT as follows. E.g.,

Constant approximation

$$f(x + h) - f(x) = \int_x^{x+h} f'(s)ds$$

Linear approximation

$$f(x + h) - f(x) - f'(x)h = \int_x^{x+h} [f'(s_2) - f'(x)]ds_2$$
$$= \int_x^{x+h} \int_x^{s_2} f''(s_1)ds_1 ds_2$$

Quadratic approximation

$$f(x+h) - f(x) - f'(x)h - \frac{h^2}{2}f'(x) = \int_x^{x+h} \int_x^{s_3} [f''(s_2) - f''(x)]ds_2 ds_3$$

$$= \int_x^{x+h} \int_x^{s_3} \int_x^{s_2} f'''(s_1)ds_1 ds_2 ds_3$$

with a remainder of repeated integral of the high derivative:

$$R(h) := \int_x^{x+h} \int_x^{s_3} \int_x^{s_2} f'''(s_1)ds_1 ds_2 ds_3,$$

$$|R(h)| \leq \underset{[x,x+h]}{\text{upper}} |f'''| \iiint_{x \leq s_1 \leq s_2 \leq s_3 \leq x+h} ds_1 ds_2 ds_3$$

$$= \underset{[x,x+h]}{\text{upper}} |f'''| \frac{1}{3!} \left(\int_x^{x+h} ds \right)^3 = \frac{h^3}{3!} \underset{[x,x+h]}{\text{upper}} |f'''|.$$

This is a special version of TT. Furthermore, each expansion can save one term (even more terms) for quadrature rules, e.g., letting $g = f'$

Trapezoid: $\int_x^{x+h} g(s_2)ds_2 - \frac{h}{2}(g(x) + g(x+h)) :\approx \frac{5h^3}{12} \underset{[x,x+h]}{\text{upper}} |g''|,$

Midpoint: $\int_x^{x+h} g(s_3)ds_3 - hg\left(x + \frac{h}{2}\right) :\approx \frac{7h^3}{24} \underset{[x,x+h]}{\text{upper}} |g''|,$

$\int_x^{x+h} g(s_4)ds_4 - hg(x) - \frac{h^2}{2}g'\left(x + \frac{h}{3}\right) :\approx \frac{5h^4}{72} \underset{[x,x+h]}{\text{upper}} |g'''|,$

$\int_x^{x+h} g(s_5)ds_5 - \frac{h}{2}(g(x) + g(x+h)) + \frac{h^3}{12}g''\left(x + \frac{h}{2}\right) :\approx \frac{19h^5}{480} \underset{[x,x+h]}{\text{upper}} |g^{(4)}|,$

Simpson $= \dfrac{2\text{Midpoint} + \text{Trapezoid}}{3} :\approx \dfrac{49h^5}{2880} \underset{[x,x+h]}{\text{upper}} |g^{(4)}|.$

(These constants can be improved). Here, we use the abbreviation $a :\approx b$ to denote $|a| \leq b$.

Remarks: Their proofs (or calculations) are long. They can be skipped for the common reader. We would like to write down for patient students. Firstly,

$$(g(x) + g(x+h)) - 2g(x) - hg'(x) = \int_x^{x+h} g'(s_3)ds_3 - hg'(x)$$

$$:\approx \frac{h^2}{2} \text{ upper } |g''|,$$
$$\underset{[x,x+h]}{}$$

$$g\left(x + \frac{h}{2}\right) - g(x) - \frac{h}{2}g'(x) = \int_x^{x+\frac{h}{2}} g'(s_2)ds_2 - \frac{h}{2}g'(x) :\approx \frac{h^2}{8} \text{ upper } |g''|,$$
$$\underset{[x,x+h]}{}$$

(where we change h to $\frac{h}{2}$). Secondly, changing g to g' and $\frac{h}{2}$ to $\frac{h}{3}$,

$$g'\left(x + \frac{h}{3}\right) - g'(x) - \frac{h}{3}g''(x) :\approx \frac{h^2}{18} \text{ upper } |g'''|.$$
$$\underset{[x,x+h]}{}$$

Thirdly,

$$(g(x) + g(x+h)) - \frac{h^2}{6}g''\left(x + \frac{h}{2}\right) - 2g(x) - hg'(x)$$

$$- \frac{h^2}{3}g''(x) - \frac{h^3}{12}g'''(x)$$

$$= \left(g(x) + g(x+h) - 2g(x) - hg'(x) - \frac{h^2}{2}g''(x) - \frac{h^3}{6}g'''(x)\right)$$

$$- \frac{h^2}{6}\left(g''(x + \frac{h}{2}) - g''(x) - \frac{h}{2}g'''(x)\right)$$

$$= \left(\int_x^{x+h}\int_x^{s_3} g''(s_4)ds_4ds_3 - \frac{h^2}{2}g''(x) - \frac{h^3}{6}g'''(x)\right)$$

$$- \frac{h^2}{6}\left(\int_x^{x+h} g'''(s_5)ds_5 - \frac{h}{2}g''(x)\right)$$

$$= \left(\int_x^{x+h}\int_x^{s_3}\int_x^{s_4} g'''(s_5)ds_5ds_4ds_3 - \frac{h^3}{6}g'''(x)\right)$$

$$- \frac{h^2}{6}\int_x^{x+h}\int_x^{s_5} g^{(4)}(s_6)ds_6ds_5$$

$$= \int_x^{x+h} \int_x^{s_3} \int_x^{s_4} \int_x^{s_5} g^{(4)}(s_6)ds_6ds_5ds_4ds_3 - \frac{h^2}{6}\int_x^{x+h}\int_x^{s_5} g^{(4)}(s_6)ds_6ds_5$$

$$:\approx \frac{h^4}{24}\underset{[x,x+h]}{\text{upper}}|g^{(4)}| + \frac{h^4}{48}\underset{[x,x+h]}{\text{upper}}|g^{(4)}| = \frac{h^4}{16}\underset{[x,x+h]}{\text{upper}}|g^{(4)}|,$$

$$\frac{2g(x+\frac{h}{2})+\frac{1}{2}(g(x)+g(x+h))}{3} - g(x) - \frac{h}{2}g'(x)$$

$$-\frac{h^2}{3!}g''(x) - \frac{h^3}{4!}g'''(x)$$

$$= \frac{2}{3}\left(g(x+\frac{h}{2}) - g(x) - \frac{h}{2}g'(x) - \frac{h^2}{8}g''(x) - \frac{h^3}{48}g'''(x)\right)$$

$$+\frac{1}{6}\left(g(x) + g(x+h) - 2g(x) - hg'(x) - \frac{h^2}{2}g''(x) - \frac{h^3}{6}g'''(x)\right)$$

$$= \frac{2}{3}\left(\int_x^{x+\frac{h}{2}}\int_x^{s_2} g''(s_3)ds_3ds_2 - \frac{h^2}{8}g''(x) - \frac{h^3}{48}g'''(x)\right)$$

$$+\frac{1}{6}\left(\int_x^{x+h}\int_x^{s_2} g''(s_4)ds_4ds_3 - \frac{h^2}{2}g''(x) - \frac{h^3}{6}g'''(x)\right)$$

$$= \frac{2}{3}\left(\int_x^{x+\frac{h}{2}}\int_x^{s_2}\int_x^{s_3} g'''(s_4)ds_4ds_3ds_2 - \frac{h^3}{48}g'''(x)\right)$$

$$+\frac{1}{6}\left(\int_x^{x+h}\int_x^{s_2}\int_x^{s_3} g'''(s_4)ds_4s_3ds_2 - \frac{h^3}{6}g'''(x)\right)$$

$$= \frac{2}{3}\int_x^{x+\frac{h}{2}}\int_x^{s_2}\int_x^{s_3}\int_x^{s_4} g^{(4)}(s_5)ds_5ds_4ds_3ds_2$$

$$+\frac{1}{6}\int_x^{x+h}\int_x^{s_2}\int_x^{s_3}\int_x^{s_4} g^{(4)}(s_5)ds_5ds_4ds_3ds_2$$

$$:\approx \frac{h^4}{576}\underset{[x,x+h]}{\text{upper}}|g^{(4)}| + \frac{h^4}{144}\underset{[x,x+h]}{\text{upper}}|g^{(4)}| = \frac{5h^4}{576}\underset{[x,x+h]}{\text{upper}}|g^{(4)}|.$$

Therefore, Trapezoid, Midpoint, Simpson, etc., hold true:

$$\int_x^{x+h} g(s_2)ds_2 - \frac{h}{2}\big(g(x) + g(x+h)\big)$$

$$= \int_x^{x+h} g(s_2)ds_2 - hg(x) - \frac{h^2}{2}g'(x) - \frac{h}{2}\big(g(x)$$

$$+g(x+h) - 2g(x) - hg'(x)\big)$$

$$:\approx \frac{h^3}{3!}\underset{[x,x+h]}{\text{upper }|g''|} + \frac{h^3}{4}\underset{[x,x+h]}{\text{upper }|g''|} = \frac{5h^3}{12}\underset{[x,x+h]}{\text{upper }|g''|},$$

$$\int_x^{x+h} g(s_3)ds_3 - hg\left(x + \frac{h}{2}\right) = \int_x^{x+h} g(s_3)ds_3 - hg(x)$$

$$-\frac{h^2}{2}g'(x) - h[g\left(x + \frac{h}{2}\right) - g(x) - \frac{h}{2}g'(x)]$$

$$:\approx \frac{h^3}{3!}\underset{[x,x+h]}{\text{upper }|g''|} + \frac{h^3}{8}\underset{[x,x+h]}{\text{upper }|g''|} = \frac{7h^3}{24}\underset{[x,x+h]}{\text{upper }|g''|},$$

$$\int_x^{x+h} g(s_4)ds_4 - hg(x) - \frac{h^2}{2}g'(x + \frac{h}{3})$$

$$= \int_x^{x+h} g(s_4)ds_4 - hg(x) - \frac{h^2}{2}g'(x) - \frac{h^3}{6}g''(x)$$

$$-\frac{h^2}{2}[g'(x + \frac{h}{3}) - g'(x) - \frac{h}{3}g''(x)]$$

$$:\approx \frac{h^4}{4!}\underset{[x,x+h]}{\text{upper }|g'''|} + \frac{h^4}{18}\underset{[x,x+h]}{\text{upper }|g'''|} = \frac{7h^4}{72}\underset{[x,x+h]}{\text{upper }|g'''|},$$

$$\int_x^{x+h} g(s_5)ds_5 - \frac{h}{2}(g(x) + g(x+h)) + \frac{h^3}{12}g''\left(x + \frac{h}{2}\right)$$

$$= \int_x^{x+h} g(s_5)ds_5 - hg(x) - \frac{h^2}{2}g'(x) - \frac{h^3}{3!}g''(x) - \frac{h^4}{4!}g'''(x)$$

$$-\frac{h}{2}\left[g(x) + g(x+h) - \frac{h^2}{6}g''\left(x + \frac{h}{2}\right) - 2g(x) - hg'(x)\right.$$

$$\left. -\frac{h^2}{3}g''(x) - \frac{h^3}{12}g'''(x)\right]$$

$$:\approx \frac{h^5}{5!}\underset{[x,x+h]}{\text{upper}}|g^{(4)}(x)| + \frac{h^5}{32}\underset{[x,x+h]}{\text{upper}}|g^{(4)}| = \frac{19h^5}{480}\underset{[x,x+h]}{\text{upper}}|g^{(4)}|,$$

$$\int_x^{x+h} g(s_2) - \frac{2g(x + \frac{h}{2}) + \frac{1}{2}(g(x) + g(x+h))}{3}$$

$$= \int_x^{x+h} g(s_2) - hg(x) - \frac{h^2}{2}g'(x) - \frac{h^3}{3!}g''(x) - \frac{h^4}{4!}g'''(x)$$

$$+ h\left(\frac{2g(x + \frac{h}{2}) + \frac{1}{2}(g(x) + g(x+h))}{3}\right) - g(x) - \frac{h}{2}g'(x)$$

$$- \frac{h^2}{3!}g''(x) - \frac{h^3}{4!}g'''(x))$$

$$:\approx \frac{h^5}{5!}\underset{[x,x+h]}{\text{upper}}|g^{(4)}| + \frac{5h^5}{576}\underset{[x,x+h]}{\text{upper}}|g^{(4)}| = \frac{49h^5}{2880}\underset{[x,x+h]}{\text{upper}}|g^{(4)}|.$$

Direct proof of Bramble-Hilbert-Xu theorem.

First, we rewrite the above TT with the function f over the interval $[x_0, x_1]$:

$$f(x) = f(x_0) + \int_{x_0}^x f'(s_1)ds_1$$

$$= f(x_0) + \int_{x_0}^x f'(x_0)ds_1 + \int_{x_0}^x \int_{x_0}^{s_1} f''(s_2)ds_2ds_1$$

$$= f(x_0) + \int_{x_0}^x f'(x_0)ds_1 + \int_{x_0}^x \int_{x_0}^{s_1} f''(x_0)ds_2ds_1$$

$$+ \int_{x_0}^x \int_{x_0}^{s_1} \int_{x_0}^{s_2} f'''(s_3)ds_3ds_2ds_1$$

$$= f(x_0) + (x - x_0)f'(x_0) + \frac{(x - x_0)^2}{2}f''(x_0)$$

$$+ \int_{x_0}^x \int_{x_0}^{s_1} \int_{x_0}^{s_2} f'''(s_3)ds_3ds_2ds_1$$

$$= f_T(x) + R(x)$$

with the Taylor polynomial $f_T(x) = f(x_0) + (x - x_0)f'(x_0) + \frac{(x-x_0)^2}{2}f''(x_0)$, and the remainder function

$$R(x) = \int_{x_0}^{x} \int_{x_0}^{s_1} \int_{x_0}^{s_2} f'''(s_3) ds_3 ds_2 ds_1, \quad |R| \le \frac{(x_1 - x_0)^3}{6} \text{ upper } |f'''|$$

and its derivatives

$$R'(x) = \int_{x_0}^{x} \int_{x_0}^{s_2} f'''(s_3) ds_3 ds_2, \quad |R'| \le \frac{(x_1 - x_0)^2}{2} \text{ upper } |f'''|,$$

$$R''(x) = \int_{x_0}^{x} f'''(s_3) ds_3, \quad |R''| \le (x_1 - x_0) \text{ upper } |f'''|,$$

$$R'''(x) = f'''(x).$$

Similar estimates also hold in the integral sense (if $|f'''|$ is integrable):

$$\int_{x_0}^{x_1} |R| dx = \int_{x_0}^{x_1} | \int_{x_0}^{x} \int_{x_0}^{s_1} \int_{x_0}^{s_2} f'''(s_3) ds_3 ds_2 ds_1 | dx$$

$$\le \int_{x_0}^{x_1} \int_{x_0}^{x_1} \int_{x_0}^{x_1} \int_{x_0}^{x_1} |f'''(s_3)| ds_3 ds_2 ds_1 dx$$

$$= (x_1 - x_0)^3 \int_{x_0}^{x_1} |f'''(s_3)| ds_3,$$

$$\int_{x_0}^{x_1} |R'| dx = \int_{x_0}^{x_1} | \int_{x_0}^{x} \int_{x_0}^{s_2} f'''(s_3) ds_3 ds_2 | dx$$

$$\le (x_1 - x_0)^2 \int_{x_0}^{x_1} |f'''(s_3)| ds_3,$$

$$\int_{x_0}^{x_1} |R''| dx = \int_{x_0}^{x_1} | \int_{x_0}^{x} f'''(s_3) ds_3 | dx \le (x_1 - x_0) \int_{x_0}^{x_1} |f'''(s_3)| ds_3,$$

$$\int_{x_0}^{x_1} |R'''| dx = \int_{x_0}^{x_1} |f'''| dx.$$

This is the B-H-× theorem[4] in one dimension with an integral norm.

Problem. B-H-× theorem works for the remainder R, what about the function f itself? E.g., for $x \in [0, 1]$, we have

$$\int_0^1 f(s_2)ds_2 - f(x) = \int_0^1 [f(s_2) - f(0)]ds_2 - [f(x) - f(0)]$$

$$= \int_0^1 \int_0^{s_2} f'(s_1)ds_1 ds_2 - \int_0^x f'(s_2)ds_2,$$

$$f(x) = \int_0^1 f(s_2)ds_2 - [\int_0^1 f(s_2)ds_2 - f(x)]$$

$$= \int_0^1 f(s_2)ds_2 + \int_0^x f'(s_2)ds_2 - \int_0^1 \int_0^{s_2} f'(s_1)ds_1 ds_2,$$

$$|f(x)| \le 2 \int_0^1 (|f(s)| + |f'(s)|)ds.$$

This is a special case of Sobolev inequality.

This means that we used low order derivatives to get a higher order error, which is called a superconvergence phenomenon.

Uniqueness of the derivative definition. We now go to answer a remainder problem for the derivative definition: f' is uniquely defined by the inequality (1.1). Otherwise, if we have two derivatives f' and f^*: For all x and nearby points $x + h$ in $[a, b]$

$$|f(x + h) - f(x) - f'(x)h| \le \epsilon(h)h$$
$$|f(x + h) - f(x) - f^*(x)h| \le \epsilon(h)h$$

then $|f'(x) - f^*(x)| \le 2\epsilon(h)$. Since $f' \not\equiv f^*$, there exists a point x_0:

$$0 < d = |f'(x_0) - f^*(x_0)| \le 2\epsilon(h) < d$$

since $\epsilon(h)$ here can be chosen small for a small h. This leads to a contradiction.

Translating derivative into the quotient of two differentials. If we denote the differential in (1.1) with $df = f'(x)h$, then $dx = h$ (where we take $f = x$, so $f' = 1$) or

$$df = f'(x)dx, \quad \frac{df}{dx} = f'(x). \tag{1.6}$$

1.2. Generalized First Inequality

Intuitively, within each local triangle over the subinterval $[x, x + h]$ the slope at arbitrary intermediate point $\xi = x + \theta h$ $(0 \le \theta \le 1)$

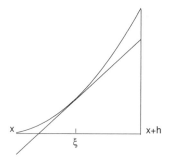

can be used to compute the height variation

$$f(x + h) - f(x) = f'(\xi)h + \text{(relative error)}h$$

where the relative error is small, see (1.8), or simply

$$|f(x + h) - f(x) - f'(\xi)h| \le \epsilon^*(h)h, \tag{1.7}$$

where $\epsilon^*(h) \approx \epsilon(h)$, independent of x, see (1.9) for details. This is called the generalized first inequality. Without making more assumptions, (1.7) can be derived by the original 1st inequality (1.1) as follows.

Indeed, ξ can be regarded as an adding node within two subintervals $[\xi, x + h]$ and $[x, \xi]$ (with different widths). Then

$$f(x + h) - f(x) = [f(x + h) - f(\xi)] + [f(\xi) - f(x)].$$

Using the right differential (1.1) and the left differential [replacing h by $-h$ in (1.1)].

$$f(x+h) - f(x) - f'(x)h \equiv \epsilon(x, h) \cdot h, \quad |\epsilon(x, h)| \leq \epsilon(h),$$
$$f(x) - f(x-h) - f'(x)h \equiv \epsilon(x, -h) \cdot h, \quad |\epsilon(x, -h)| \leq \epsilon(-h),$$

we have

$$f(x+h) - f(\xi) = f'(\xi)(1-\theta)h + \epsilon(\xi, (1-\theta)h)(1-\theta)h,$$
$$|\epsilon(\xi, (1-\theta)h)| \leq \epsilon((1-\theta)h),$$
$$f(\xi) - f(x) = f'(\xi)(\theta h) + \epsilon(\xi, -\theta h)(\theta h),$$
$$|\epsilon(\xi, -\theta h)| \leq \epsilon(-\theta h).$$

Then

$$f(x+h) - f(x) = f'(\xi)h + \overline{\epsilon}(\xi, x, h)h, \qquad (1.8)$$

$$\overline{\epsilon}(\xi, x, h)h = \epsilon(\xi, (1-\theta)h)(1-\theta)h + \epsilon(\xi, -\theta h)(\theta h).$$

or simply

$$|\overline{\epsilon}(\xi, x, h)| \leq \epsilon^*(h) \equiv \max_{0 \leq \theta \leq 1} [\epsilon((1-\theta)h), \quad \epsilon(-\theta h)]. \qquad (1.9)$$

By combining (1.8) and (1.9) we get (1.7).

1.3. Generalized Second Inequality

Since the first inequality, (1.1), has a more general form, (1.8), or simply

$$|f(x+h) - f(x) - f'(\xi)h| \leq \epsilon^*(h)h, \quad \xi \in [x, x+h],$$

adding up these inequalities on each subinterval $[x, x+h]$ (where the width h is the same)

the generalized second inequality

$$|f(b) - f(a) - \sum_{\xi \in [x, x+h]} f'(\xi)h| \leq (b-a)\epsilon^*(h) \qquad (1.10)$$

is obtained.

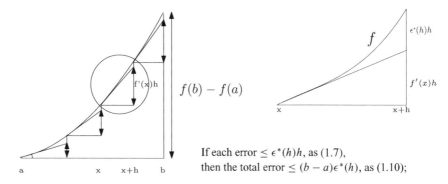

If each error $\leq \epsilon^*(h)h$, as (1.7),
then the total error $\leq (b-a)\epsilon^*(h)$, as (1.10);

that is why we define the differential as (1.7).

Let us give the detailed proof of (1.10) as the following familiar arguments.

Let the total interval $[a, b]$ be divided, for simplicity, into $n + 1$-equal subintervals $[x_i, x_i + h]$, where $a = x_0 < x_1 < \cdots < x_{n+1} = b$, $h = x_{i+1} - x_i$.

Use the general inequality (1.8) to each subinterval:

$$f(x_0 + h) - f(x_0) - f'(\xi_0)h = \bar{\epsilon}(\xi_0, x_0, h) \cdot h$$
$$f(x_1 + h) - f(x_1) - f'(\xi_1)h = \bar{\epsilon}(\xi_1, x_1, h) \cdot h$$
$$\cdots = \cdots$$
$$f(x_n + h) - f(x_n) - f'(\xi_n)h = \bar{\epsilon}(\xi_n, x_n, h) \cdot h,$$

where the error terms on the right-hand side satisfy

$$|\bar{\epsilon}(\xi_i, x_i, h)| \leq \epsilon^*(h), \quad 0 \leq i \leq n,$$

and add up. The terms on the left-hand side add up to give

$$f(b) - f(a) - \sum_{i=0}^{n} f'(\xi_i)h,$$

and the total error is the average of relative error in (1.8):

$$\left| \sum_{i=0}^{n} \bar{\epsilon}(\xi_i, x_i, h) \cdot h \right| \leq (b-a)\epsilon^*(h)$$

or (1.10).

A similar argument can be used for the unequal subintervals $[x_i, x_i + h_i]$.

1.4. Rules of Differentiation

As mentioned at the end of Sec. 1.1 that the derivative's definition itself is constructive, any elementary function in principle can be calculated from the definition itself. However, we can use other rules that will make the process easier. Indeed, from the derivatives of the three elementary test functions in Sec. 1.1, we can get arithmetics of derivatives for elementary functions and derivatives of composite functions and inverse functions. We borrow the words and proofs used in this section from Strang's *Calculus* (Sec. 2.5). If readers, however, are interested only on the theoretical principle, the following proofs (even are mechanically) can be skipped, just believe the rules.

1.4.1. *Arithmetic of derivatives*

In this section we will provide the following arithmetics:

(i) Sum rule: If f and g are uniformly differentiable on $[a, b]$, then $f + g$ is also uniformly differentiable on $[a, b]$, and the derivative obeys the sum rule $(f + g)' = f' + g'$. This result is what we hope for: When we add functions, we add their derivatives.

(ii) Rule of linearity: If f is uniformly differentiable on $[a, b]$, then cf is also uniformly differentiable on $[a, b]$, and the derivative obeys the rule of scalar product $(cf)' = cf'$. This result is again what we hope for.

(iii) Product rule: If f and g are uniformly differentiable on $[a, b]$, then fg is also uniformly differentiable on $[a, b]$, and the derivative obeys the product rule $(fg)' = f'g + fg'$. This is not so simple as we desired. The derivative of f times g is not f' times g'.

(iv) Reciprocal rule: If f is uniformly differentiable on $[a, b]$ and $|f(x)| \geq m > 0$, then $\frac{1}{f}$ is also uniformly differentiable on $[a, b]$, and the derivative obeys the reciprocal rule $(\frac{1}{f})' = -\frac{f'}{f^2}$.

(v) Quotient rule: If f and g are uniformly differentiable on $[a, b]$ and $|f(x)| \geq m > 0$, then $\frac{g}{f}$ is also uniformly differentiable on $[a, b]$, and the derivative obeys the quotient rule $(\frac{g}{f})' = -\frac{f'g - fg'}{f^2}$.

(vi) Power rule: The derivative of $[f(x)]^n$ is $n[f(x)]^{n-1} f'$.

Proof. It is evident that the rule of linearity, (i) and (ii), holds. (vi) is repeated applications of (iii).

For saving notations and calculations we only check the simplest uniform differentiability (1.1) and (1.5):

$$f(x + h) - f(x) - f'(x)h = O(h^2),$$
$$|O(h^2)| \leq Ch^2.$$

The proof of the product rule (iii):

$$
\begin{aligned}
f(x + h)&g(x + h) - f(x)g(x) \\
&= f(x + h)g(x + h) - f(x + h)g(x) + f(x + h)g(x) - f(x)g(x) \\
&= f(x + h)(g(x + h) - g(x)) + g(x)(f(x + h) - f(x)) \\
&= f(x)(g(x + h) - g(x)) + g(x)(f(x + h) - f(x)) \\
&\quad + (f(x + h) - f(x))(g(x + h) - g(x)) \\
&= f(x)(g'(x)h + O(h^2)) + g(x)(f'(x)h + O(h^2)) \\
&\quad + (f'(x)h + O(h^2))(g'(x)h + O(h^2)) \\
&= (f(x)g'(x) + f'(x)g(x))h + O(h^2).
\end{aligned}
$$

Therefore, fg is uniformly differentiable on $[a, b]$ and $(fg)' = f'g + fg'$.

The proof of the reciprocal rule (iv) starts with $(f)(1/f) = 1$. The derivative of 1 is 0. Apply the product rule:

$$f\left(\frac{1}{f}\right)' + \frac{1}{f}f' = 0 \quad \text{so that} \quad \left(\frac{1}{f}\right)' = \frac{-f'}{f^2}.$$

It can also be proved by the original definition of the derivative:

$$\frac{1}{f(x+h)} - \frac{1}{f(x)} = \frac{f(x) - f(x+h)}{f(x+h)f(x)}$$

$$= \frac{-f'(x)h + O(h^2)}{f(x+h)f(x)}$$

$$= \frac{-f'(x)h + O(h^2)}{(f(x))^2} + \frac{-f'(x)h + O(h^2)}{f(x+h)f(x)}$$

$$- \frac{-f(x)'h + O(h^2)}{(f(x))^2}$$

$$= \left(-\frac{f'(x)}{(f(x))^2}\right)h + \frac{O(h^2)}{(f(x))^2}$$

$$+ \left(\frac{-f(x)'h + O(h^2)}{f(x)}\right)\left(\frac{1}{f(x+h)} - \frac{1}{f(x)}\right)$$

$$= \left(-\frac{f'(x)}{(f(x))^2}\right)h + O(h^2) + O(h)\frac{f(x) - f(x+h)}{f(x+h)f(x)}$$

$$= \left(-\frac{f'(x)}{(f(x))^2}\right)h + O(h^2).$$

Thus $\frac{1}{f}$ is uniformly differentiable on $[a, b]$ and $(\frac{1}{f})' = -\frac{f'}{f^2}$.

Combining the product rule (iii) with the reciprocal rule (iv), we obtain the quotient rule (v): If f and g satisfy the uniform inequalities (1.1) and (1.5) on $[a, b]$, then $\frac{g}{f}$ is uniformly differentiable on $[a, b]$ and $(\frac{g}{f})' = -\frac{f'g - fg'}{f^2}$.

The proof of the power rule (vi): For $n = 1$ this reduces to $f' = f'$. For $n = 2$ we get the square rule $2ff'$. Next comes f^3. The best approach is to use mathematical induction, which goes from each n to the next power

$n + 1$ by the product rule:

$$(f^{n+1})' = (f^n f)' = f^n f' + f(nf^{n-1} f') = (n + 1) f^n f'.$$

That is exactly rule (vi) for the power $n + 1$. We get all positive powers this way, going up from $n = 1$, then the negative powers from the reciprocal rule. □

1.4.2. Derivatives of rational functions and trigonometric functions

(i) From $(x)' = 1$ and using the product rule (iii) and mathematical induction, we see that x^n is uniformly differentiable and $(x^n)' = nx^{n-1}$.

(ii) Polynomial functions are uniformly differentiable and

$$(a_0 + a_1 x + a_2 x^2 + \cdots + a_n x^n)' = a_1 + 2a_2 x + 3a_3 x^2 + \cdots + na_n x^{n-1}.$$

(iii) From $(x)' = 1$ and using the reciprocal rule (iv) and mathematical induction, we see that $\frac{1}{x^n}$ is uniformly differentiable on any closed interval that do not contain 0 and

$$\left(\frac{1}{x^n}\right)' = \frac{-n}{x^{n+1}} \quad \text{or} \quad (x^{-n})' = -nx^{-n-1}.$$

The interesting thing is that this answer $-nx^{-n-1}$ fits into the same pattern as x^n in (i). Multiply by the exponent and reduce it by one.

So, the power rule, $(x^n)' = nx^{n-1}$, applies when n is positive or negative, but n must be a whole number.

(iv) We want to allow fractional powers $n = p/q$ and keep the same formula. The derivative of x^n is still nx^{n-1}: Write $f = x^{p/q}$ as $f^q = x^p$. Take derivatives, assuming they exist:

$$qf^{q-1} f' = px^{p-1} \quad \text{[power rule (vi) for both sides]}$$
$$f' = \frac{px^{-1}}{qf^{-1}} \quad \text{[cancel } x^p \text{ with } f^q]$$
$$f' = nx^{n-1} \quad \text{[replace } p/q \text{ by } n \text{ and } f \text{ by } x^n]$$

So, the power rule, $(x^n)' = nx^{n-1}$, applies when n is negative, or a fraction (or even any real number).

(v) Rational functions satisfy the uniform inequalities (1.1) and (1.5) on any closed interval in which the denominator is not equal to 0. We use the rule of linearity, (i) and (ii), and the quotient rule (v) to get the derivatives.

(vi) We know from Sec. 1.1 that the uniform derivatives of $\sin x$ and $\cos x$ are known. Using the quotient rule (v), we can get the uniform derivatives of the tangent and the cotangent:

$$(\tan x)' = \sec^2 x$$
$$(\text{ctg}x)' = -\csc^2 x.$$

By the reciprocal rule (iv) we can get the uniform derivatives of the secant and the cosecant:

$$(\sec x)' = \sec x \tan x,$$
$$(\csc x)' = -\csc x \text{ctg}x.$$

These formulas can be easily proved.

1.4.3. Derivatives of composite functions and inverse functions

(i) Chain rule: If $f(x)$ is uniformly differentiable on $[a, b]$, the range of $f(x)$ is no more than $[u, v]$, and $g(x)$ is uniformly differentiable on $[u, v]$, then the composite function $g(f(x))$ is uniformly differentiable on $[a, b]$ and

$$(g[f(x)])' = g'[f(x)]f'(x).$$

This is power rule when $g(y) = y^n$. The proof is as follows:

$$\begin{aligned}
g(f(x+h)) - g(f(x)) &= g'(f(x))(f(x+h) - f(x)) + O(h^2) \\
&= g'(f(x))(f'(x)h + O(h^2)) + O(h^2) \\
&= g'(f(x))f'(x)h + g'(f(x))O(h^2) + O(h^2) \\
&= g'(f(x))f'(x)h + O(h^2).
\end{aligned}$$

We complete the proof.

Chain rule is the hardest rule of differentiation. First, we must make the composition of the function clear.

Examples:

$$y = \sin x^2 \text{ is rewritten as } y = \sin z, \quad z = x^2;$$
$$y = \sin^2 x \text{ is rewritten as } y = z^2, \quad z = \sin x;$$
$$y = \sqrt{1 - x^2} \text{ is rewritten as } y = z^{\frac{1}{2}}, \quad z = 1 - x^2.$$

(ii) Derivative of inverse function: If $f(x)$ is uniformly differentiable on $[a, b]$ and the derivative satisfies $f' \geq m > 0$, then we see from the FT that f is strictly increasing. Therefore, f has the inverse function $G(x)$ on $[f(a), f(b)]$. Then we will prove $G(x)$ is uniformly differentiable and

$$G'(x) = \frac{1}{f'(G(x))} \quad (f(G(x)) = x).$$

The proof is as follows: From $f(G(x)) = x$ the chain rule gives $f'(G(x))G'(x) = 1$. Writing $y = G(x)$ and $x = f(y)$, this rule looks better:

$$\frac{dx}{dy}\frac{dy}{dx} = 1 \quad \text{or} \quad \frac{dx}{dy} = \frac{1}{dy/dx}.$$

The slope of $x = G^{-1}(y)$ times the slope of $y = G(x)$ equals 1.

Examples:

If $y = e^x$, $x = \ln y$, then from (iii) and (1.6) of Sec. 1.1, we have:

$$\frac{dy}{dx} = y' = e^x;$$

therefore

$$\frac{dx}{dy} = \frac{1}{e^x} = \frac{1}{y} \quad \text{or} \quad (\ln y)' = \frac{1}{y};$$

If $y = \sin x$, $x = \sin^{-1} y$, then

$$\frac{dy}{dx} = y' = \cos x,$$

$$\frac{dx}{dy} = \frac{1}{\cos x} = \frac{1}{\sqrt{1 - \sin^2 x}} = \frac{1}{\sqrt{1 - y^2}}.$$

Here are the six functions for quick reference:

$$\begin{array}{ll}
\text{function } f(y) & \text{slope } dx/dy \\[4pt]
\sin^{-1} y, \cos^{-1} y & \pm \dfrac{1}{\sqrt{1 - y^2}} \\[10pt]
\tan^{-1} y, \cot^{-1} y & \pm \dfrac{1}{1 + y^2} \\[10pt]
\sec^{-1} y, \csc^{-1} y & \pm \dfrac{1}{|y|\sqrt{y^2 - 1}}
\end{array}$$

The column of derivative is what we need and use in calculus.

1.5. Tables of Derivatives and Integrals

Calculus relies heavily on a very few functions, simple elementary functions, including the polynomial, the sine and cosine, the ln, and exponential functions (see Strang's *Calculus*).[25] Fortunately, the definition of derivative, (1.1), itself has been used to compute the derivatives of these functions: For example from Secs. 1.1 and 1.4,

$$(x^n)' = nx^{n-1}$$
$$(\sin x)' = \cos x,$$
$$(\cos x)' = -\sin x,$$
$$(e^x)' = e^x,$$
$$(\ln x)' = \frac{1}{x}.$$

Conversely, integration (1.3) gives the integral table: Each integration formula came directly from a differentiation formula, e.g.,

$$\int_a^b x^n dx = \frac{b^{n+1}}{n+1} - \frac{a^{n+1}}{n+1} \quad (n \neq -1),$$

$$\int_a^b \cos x\, dx = \sin b - \sin a,$$

$$\int_a^b \sin x\, dx = -\cos b + \cos a,$$

$$\int_a^b \frac{1}{x}dx = \ln b - \ln a,$$

$$\int_a^b e^x dx = e^b - e^a.$$

The logarithm fills a gap among the integrals, $n = -1$.

1.6. Rules of Integration

(i) Integration by parts: Let $f' = u$ and $g' = v$. From the FT, (1.3),

$$\int_a^b u dx = f(b) - f(a) \text{ and } \int_a^b v dx = g(b) - g(a)$$

we have the sum rule:

$$\int_a^b (u + v)dx = \int_a^b (f + g)'dx = f(b) + g(b) - f(a) - g(a)$$
$$= \int_a^b u dx + \int_a^b v dx.$$

From the FT and the product rule:

$$u(b)v(b) - u(a)v(a) = \int_a^b (uv)'dx = \int_a^b (uv' + u'v)dx$$
$$= \int_a^b uv' dx + \int_a^b u' v dx$$

we have integration by parts:

$$\int_a^b uv' dx = u(b)v(b) - u(a)v(a) - \int_a^b u' v dx.$$

Integration by parts is a small trick (in deciding which one is u or v).

For simplicity, we ignore the lower and upper limits, a and b, below.

Examples: $\int xe^x dx = ?$ Let $u = x$ and $v' = e^x$; hence $u' = 1$, $v = e^x$, and

$$\int xe^x dx = xe^x - \int 1 \cdot e^x dx;$$

$\int \ln x dx = ?$ Let $u = \ln x$ and $v' = 1$; hence $u' = \frac{1}{x}$, $v = x$, and

$$\int \ln x dx = x\ln x - \int dx.$$

Some examples even need repeated integration by parts:

$$\int e^x \sin x dx = e^x \sin x - \int e^x \cos x dx$$

$$= e^x \sin x - e^x \cos x - \int e^x \sin x dx.$$

Integration by parts is not just a trick but is a foundation for the theory of differential equations (and even generalized functions). See Sec. 3.4.

(ii) Integration by substitution: From the chain rule (i) of Sec. 1.4.3,

$$(g[f(x)])' = v[f(x)]f'(x), \quad \text{where } v = g',$$

we have the integration by substitution:

$$\int_a^b v[f(x)]f'(x)dx = \int_a^b (g[f(x)])'dx$$

$$= g[f(b)] - g[f(a)] = \int_{f(a)}^{f(b)} v(x)dx.$$

The integration by substitution is more tricky.

Example:

$$\int e^{kx} dx = \int e^t \frac{1}{k} dt = \frac{1}{k} \int e^t dt$$

(with $kx = t$ and $kdx = dt$);

$$\int \frac{xdx}{\sqrt{a^2 - x^2}} = -\int dt$$

(with t $= \sqrt{a^2 - x^2}$ and $dt = -\frac{x\,dx}{\sqrt{a^2-x^2}}$);

$$\int \sqrt{a^2 - x^2}\,dx = \int \sqrt{a^2 - a^2 \sin^2 u}\; a \cos u\,du$$

$$= a^2 \int \cos^2 u\,du = a^2 \int \frac{1 + \cos 2u}{2}\,du$$

(with $x = a \sin u$).

Although the integration by substitution greatly extends the scope of integration still many integrals cannot be computed. Integration is much harder than differentiation.

Differentiation in Sec. 1.4 gives derivative computations for all elementary functions. On the contrary, there exist those elementary functions whose integrals are not elementary functions, e.g., e^{-x^2}, $\frac{1}{\ln x}$, $\frac{\sin x}{x}$. To get these integrals, we have recourse to numerical integration with a sufficiently small error; cf. (ii) of the following section.

1.7. A Calculus Net

Having the FT, (1.3), we follow various formulas and properties in calculus with a check and without proof, including the properties of the differentiable function, the Taylor formula and Taylor series, and the solvers of differential equations:

(i) Properties of the differentiable function: If $f' \equiv 0$, then

$$f \equiv c;$$

if $f' \geq 0$, then $f \uparrow$ (and so the derivative is indeed an interval phenomenon);

(ii) Taylor theorem, quadrature rules, B-H-× theorm and Taylor series;

(iii) Rules of integration, including the integration by substitution, integration by parts;

(iv) Solver of the simplest differential equation: If there exists a curve f whose slopes are known,

$$f' = g \tag{1.11}$$

[where f' is the uniform derivative satisfying (1.1)], then we can solve

$$f(x) = f(a) + \int_a^x g(t)dt. \tag{1.12}$$

There are other varieties, i.e., those equations that can be put into the simplest differential equation, e.g., the linear equation and separate variable equation (see Sec. 2.2 of Chap. 2), and even the second-order linear differential equation (see Sec. 3.1 of Chap. 3).

For the existence problem of (1.11) see Sec. 2.1.

(v) The FT (1.3) shows the ties between different measurements. When the height f is not known, define f from the area of f'. Conversely, when the height f is known, (1.3) gives the area of f', where the height of the hill is equal to the area enclosed by the slope curve.

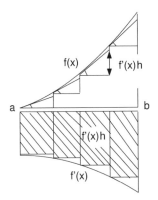

Since slopes are the most fundamental quantity of a height curve we record all of them and get a slope curve, an inverted image of the height curve.

(vi) Arclength measurement: Besides the height and area measurements, the above figure also contains an arclength measurement,

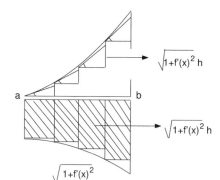

where

$$\text{arclength} = \int_a^b \sqrt{1 + f'(x)^2}dx.$$

See Sec. 0.4 in the last chapter.

These materials contain all creams in calculus and they are stringed into the FT.

So, calculus is so simple: Simply remember a definition of the differential and their sum (over all subintervals)—the FT.

1.8. Taylor's Series

From the TT in Sec. 1.1, we have

$$f(x) - f(\phi) = f'(\phi)x + \cdots + \frac{1}{n!}x^n f^{(n)}(\phi) + R,$$

$$|R| \le \frac{x^{n+1}}{(n+1)!} \text{ upper } |f^{(n+1)}|.$$

If $f^{(n+1)}$ is bounded after $n > N$, the whole infinite sum is a Taylor series:

$$e^x = 1 + x + \frac{x^2}{2!} + \frac{x^3}{3!} + \cdots,$$

$$\cos x = 1 - \frac{x^2}{2!} + \frac{x^4}{4!} - \frac{x^6}{6!} + \cdots,$$

$$\sin x = x - \frac{x^3}{3!} + \frac{x^5}{5!} - \frac{x^7}{7!} + \cdots.$$

These series are the base in solving oscillating differential equations in Sec. 3.1 of Chap. 3.

1.9. Euler's Formula

Define e^{ix} as a series

$$e^{ix} = \sum_{j=0}^{\infty} \frac{1}{j!}(ix)^j.$$

Using

$$i^2 = -1$$

and splitting the series into even terms and odd terms we get

$$e^{ix} = \sum_{j=0}^{\infty} \frac{(-1)^j}{(2j)!}(x)^{2j} + i\sum_{j=0}^{\infty} \frac{(-1)^j}{(2j+1)!}(x)^{2j+1}$$
$$= \cos x + i \sin x.$$

See Strang's *Calculus*, p. 389.[25]

Furthermore,

$$(e^{ix})' = -\sin x + i \cos x = i(\cos x + i \sin x) = ie^{ix}.$$

For the same reason, for a complex number λ,

$$(e^{\lambda x})' = \lambda e^{\lambda x}, \quad (e^{\lambda x})'' = \lambda^2 e^{\lambda x},$$

etc., i.e., the derivative of such a function, $e^{\lambda t}$, keeps the same type (different by a multiple), which will be used in solving differential equations like $au'' + bu' + cu = 0$ in Sec. 3.1 of Chap. 3.

Integration by parts is a foundation for the theory of differential equations. See Sec. 3.2 of Chap. 3.

We have arrived at our aim, or the first phase calculus, with nothing hard or long. Indeed, put aside everything but remember a definition of the differential and their sum (over all subintervals)—the FT.

For the next phase we want to know more, e.g., the existence theorem of the differential equation (1.11) for any continuous function f and more differential equations.

1.10. Possible Generalizations

All formulas (1.1) to (1.12) hold also for an abstract function $f(x)$ in a general linear norm space, with the independent variable x in $[a, b]$ (e.g., a vector function of the real variable). We recall that the original proof of the FT in Ljusternik–Sobolev's Elements of Functional Analysis (Sec. 1 of Chap. 8)[22] is based on the Hahn–Banach theorem (and the FT for the real-valued function); so the proof is very long, even not complete. In this booklet, the FT is given by the sum of definitions (1.1) themselves without proof (or proved in a few words). It greatly simplifies the original treatment in Ljusternik–Sobolev's Elements of Functional Analysis.[22]

Although the dependent variable f can be easily generalized into a linear norm space, there is no obvious counterpart to the higher-dimensional independent variable, e.g., $f(x, y)$. So far, only some special cases have been examined. See the following examples.

(i) Green's formula

Consider the two-variable functions

$$f(x, y), \quad g(x, y)$$

defined on a rectangle $D = [a, b] \times [c, d]$ with the boundary ∂D^+.

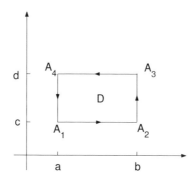

Green's formula under the uniform definition (1.1): Let f and g be uniformly partial differentiable with respect to x and y, respectively. Then

$$\oint_{\partial D^+} f(x, y)dx + g(x, y)dy = \iint_D \left(\frac{\partial g}{\partial x} - \frac{\partial f}{\partial y} \right) dxdy.$$

Proof. This two-dimensional FT is reduced into the one dimension:

$$\iint_D \frac{\partial g}{\partial x} dxdy = \int_c^d \left[\int_a^b \frac{\partial g}{\partial x} dx \right] dy$$

$$= \int_c^d [g(b, y) - g(a, y)]dy$$

$$= \int_c^d g(b, y)dy - \int_c^d g(a, y)dy$$

$$= \int_{\overline{A_2 A_3}} g(x, y)dy + \int_{\overline{A_4 A_1}} g(x, y)dy$$

$$= \int_{\overline{A_1 A_2} + \overline{A_2 A_3} + \overline{A_3 A_4} + \overline{A_4 A_1}} g(x, y)dy$$

$$= \oint_{\partial D^+} g(x, y)dy,$$

i.e.,

$$\oint_{\partial D^+} gdy = \iint_D \frac{\partial g}{\partial x} dxdy.$$

In the same way

$$\oint_{\partial D^+} fdx = - \iint_D \frac{\partial f}{\partial y} dxdy.$$

$\square$

(ii) The height formula

If a surface is denoted by a two-variable function $f(x, y)$ defined on the rectangle $[a, x; b, y]$, then the height of the surface at point P, Q can be decomposed by

$$f(Q) - f(P) = f(R) - f(P) + f(Q) - f(R),$$

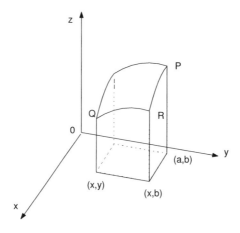

i.e., the sum of height variations in horizonal and vertical directions. Thus we can use the one-variable FT to get

$$f(Q) - f(P) = \int_a^x f_x(x, b)dx + \int_b^y f_y(x, y)dy.$$

This is a two-variable integral formula.

(iii) The area formula

Again, by the infinitesimal method, take out a small surface, which is replaced by a tangent plane:

$$S_{\text{ABCD}} \approx S_{\text{AMPN}},$$

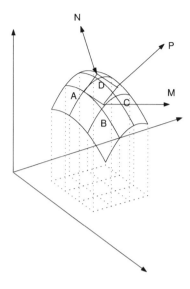

where we use the tangent instead of the secant because the former is simpler than the latter.

The above graph is too complex; let us enlarge the graph of small surfaces horizontally and vertically and add with corresponding coordinates.

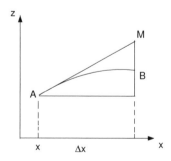
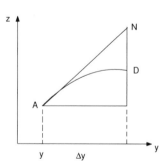

We make it simpler: let AMPN be a parallelogram. We then compute the area of AMPN in the three-dimensional space:

$$S_{\mathrm{AMPN}} = \sqrt{1 + f_x^2 + f_y^2} \cdot \Delta x \, \Delta y.$$

Adding these areas, we get

$$S = \int \int_D \sqrt{1 + f_x^2 + f_y^2} \, dx \, dy.$$

There are only three simple examples. Many formulas of multivariable calculus are hard to understand.

Chapter 2

Differential Equations of First Order

Outline: Differential equations = varieties of FT

2.1. A Simplest Differential Equation

Let us return to the simplest differential equation (1.11):

$$f' = g, \quad f(a) = \alpha.$$

An existence problem: For a given right-hand side g, is there a solution f satisfying (1.11)? We can prove by a constructive approach from Johson *et al.*'s *Computational Differential Equations* (pp. 39–49) or J. Liu's manuscript that if g is continuous (uniformly):

$$\underset{x}{\text{upper}} \, |g(x + h) - g(x)| \leq \epsilon(h)$$

with a small $\epsilon(h)$, then there exists a unique solution, (1.12):

$$f(x) = f(a) + \int_a^x g(s)ds$$

satisfing the differential equation (1.11). Indeed, what we can do is construct special functions, e.g., constant functions or linear functions (in the piecewise sense) g_N that approach g:

$$\|g_N - g\| \leq \epsilon\left(\frac{1}{N}\right)$$

(where we use the maximum norm: $\|u\| = \max_x |u(x)|$, cf. Chap. 4), and that we can directly calculate their primitive functions f_N:

$$f'_N = g_N, \quad f_N(a) = \alpha,$$

$$\|f'_N - f'_M\| \leq \epsilon\left(\frac{1}{N}\right) + \epsilon\left(\frac{1}{M}\right), \quad f_N(a) - f_M(a) = 0,$$

$$\|f_N - f_M\| \leq \epsilon\left(\frac{1}{N}\right) + \epsilon\left(\frac{1}{M}\right).$$

Then, intuitively, there exists a function f such that

$$\|f_N - f\| \leq \epsilon\left(\frac{1}{N}\right),$$

where

$$f' = g.$$

We complete the proof of the existence theorem.

So, the uniform derivative (1.1) completely tallies with the differential equation.

2.2. Varieties of Simplest Differential Equation

Besides the simplest differential equation (1.11), we will face other types of differential equation, in which the derivative is explained as a velocity.

2.2.1. The test equation

Consider the type

$$f'(x) = cf(x),$$

the most important differential equation in applied mathematics (see Strang's *Calculus*, p. 242)[25] that all analysts use. This step is a breakthrough since we cannot obtain the solution $f(x)$ from the FT directly.

A heuristic solution is to rewrite the differential using (1.6):

$$df = f'(x)dx,$$

then

$$\frac{df}{f} = cdx,$$

and the integral is

$$\ln f = cx + c_1.$$

Then f itself is $f = c_2e^{cx}$ (see Strang's *Calculus*, p. 242),[25] which can now be checked directly—discovering first and proving afterwards. This new differential equation is actually a variety of the simplest differential equation.

Another rigorous but clever solution, suggested by J. Liu is to transform the equation into

$$e^{-rx}f' = e^{-rx}rf'$$

Hence

$$(e^{-rx}f')' = \phi \quad \text{or} \quad e^{-rx}f = c \quad \text{or} \quad f = ce^{rx}$$

This differential equation has a real-life application in China in recent years. With such a differential equation we can assess the population of China in 2000: There was a census—a door to door operation throughout the entire nation that took more than a year—projecting a population of 12.66 hundred million, whereas a population assessment can be made by a college student indirectly by solving a differential equation in a few minutes, giving 13.45 hundred million as the country population (a difference of 80 million, covering nearly 6.5 percent). The two results are almost the same but their made of arrival makes a world of difference. That is why Newton said: "It is necessary to solve differential equations."

How does the student get such a number in a few minutes?

Let $f(x)$ express the population number at time x. There is a population equation $f'(x) = cf(x)$ with the growth rate $c = 0.0148$, and a known data $f(1990) = 11.6$ (hundred million), see, e.g., Zhu–Zheng's *Calculus*.[31] Finally, by the solution $f = c_2e^{cx}$ mentioned above, $f(2000) = 13.45$ (hundred million).

Furthermore, we can solve more differential equations that can be put in the form of the simplest differential equation.

2.2.2. Linear differential equation

Consider the linear differential equation of the type

$$f' + \alpha(x)f = g.$$

Multiply it by $\beta(x)$:

$$\beta f' + \alpha(x)\beta f = \beta g,$$

and modify the left-hand side to a derivative form:

$$\begin{cases} (\beta f)' = \beta f' + \beta' f, \\ \beta' = \alpha(x)\beta. \end{cases}$$

Then we can solve

$$\begin{cases} \beta f = \int (\beta f)' dx + c = \int \beta g \, dx + c, \\ \beta(x) = ce^{\int \alpha(x)dx}. \end{cases}$$

2.2.3. Separable equation

Consider the separable equation of the type

$$f' = \frac{u(x)}{v(f)}.$$

Multiply it by $v(f)$:

$$v(f)f' = u(x),$$

and modify the left-hand side to a derivative form:

$$\begin{cases} (G[f(x)])' = G'(f)f', \\ G'(f) = v(f). \end{cases}$$

Then we can solve

$$\begin{cases} G[f(x)] = \displaystyle\int u(x)dx + c, \\ G(f) = \displaystyle\int v(f)dx. \end{cases}$$

Remark. Or simply rewrite $df = f'(x)dx$ for the differential, then

$$u(x)dx = v(f)df,$$

and the integral is

$$\int u(x)dx = \int v(f)df.$$

Then you could check by differentiation that f satisfies the equation.

Those differential equations, including the exact equation, together with many real-life examples, can be found in Braun's *Differential Equations and Their Applications*.[3]

2.3. More General Equations

We have solved in the last section the differential equation of the type $f' = g(f)$. For the general first-order differential equations

$$f' = g(x, f)$$

we cannot expect an explicit solver. Mathematics has to be manipulated to answer the existence and uniqueness theorem. See Braun's *Differential Equations and Their Applications*.[3]

We now turn to the approximate method, Euler's algorithm tested with the simplest differential equation and its variety.

2.4. Tests for Euler's Algorithm

This is the predecessor of the FT (before limits):

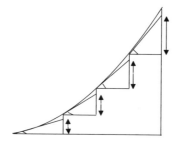

But the discontinuous broken line here is replaced by a continuous broken line through moving it parallelly,

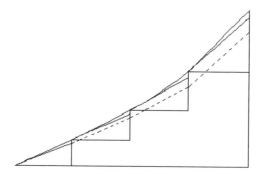

where the two calculations (continuous and discontinuous) for the hill height are the same. So, Euler's algorithm and the fundamental solver have essentially the same figure. Both of them are based on the differential, the height of the tangent line within the short segment. The former is a broken line (within a short segment) while the latter is a curve (within an infinitesimal segment). The exact formula and the approximate computation are unified:

$$\text{Euler's algorithm} \approx \text{FT}.$$

The error of Euler's algorithm is also the same as Sec. 1.1. The algorithm is

$$f_1 = f_0 + g(x_0)h, \quad f_{n+1} = f_n + g(x_n)h, \tag{2.1}$$

where $a = x_0 < x_1 < \cdots < x_n < x_{n+1} = b$, $h = x_{i+1} - x_i$, $f_0 = f(a)$, and $f_{n+1} \approx f(b)$. Or other

$$f_{n+1} = f(a) + \sum_{0 \leq i < n+1} g(x_i)h.$$

This is the predecessor or a finite analogy of the FT with the error

$$e_{n+1} = f(b) - f_{n+1} = f(b) - f(a) - \sum_{0 \leq i < n+1} g(x_i)h,$$

the same as the proof for TT in Chapter 1:

$$e = \int_a^b g(x)dx - \sum g(x)h$$

$$= \sum \int_x^{x+h} (g(s) - g(x))ds = \sum \int_x^{x+h} \int_x^s g'(w)dwds,$$

$$|e_{n+1}| \leq \frac{1}{2} \sum_{a \leq x < b} \left(\underset{x < w < x+h}{\text{upper}} |g'(w)| \right) h^2.$$

Thus Euler's algorithm tested with the simplest differential equation is nothing more than the FT.

We now test Euler's algorithm with the equation, $f' = cf$, that all analysts use:

$$f_{n+1} = (1 + ch)f_n = (1 + ch)^{n+1} f_0$$

$$= (1 + ch)^{\frac{c(b-a)}{ch}} f(a) \to e^{c(b-a)} f(a),$$

where we assume $(n + 1)h = b - a$, cf. Strang's *Calculus* (p. 274).[25]

2.5. General Euler's Algorithm

Equation (1.9) is the simplest differential equation where we use the integral in (1.10), $\int_a^b g(x)dx$, to define the solution and use Euler's algorithm with an explicit error bound, even the adaptive error control (cf. Eriksson–Estep–Hansbo–Johson's *Computational Differential Equations*).[9] These accurate formulas cannot be moved to the general first-order differential equations as done in Sec. 2.3:

$$f' = g(x, f),$$

but the source formula (2.1) can have an extension form

$$f_{n+1} = f_n + g(x_{n+1}, f_n)h_n, \quad f_0 = f(a)$$

with a possible convergence result when the solving interval is short enough. The latter with some varieties (e.g., the Runge-Kutta method, see Braun's *Differential Equations and Their Applications*,[3] and Strang's *Calculus*)[25] is a widely used algorithm nowadays. Hence, Strang said in his *Calculus* (p. 27)[25]: "modern mathematics is a combination of exact formulas and approximate computations. Neither part can be ignored."

Chapter 3

Differential Equations of Second Order

Outline: Euler's algorithm = FT before taking limit
Finite element algorithm = scant approximation

3.1. Initial Value Problems

In the last chapter (Secs. 2.1 and 2.2), we just solved several types of first-order differential equations. This chapter turns to the second-order differential equations containing the second derivative f''.

If f'' is small, it means the first derivative f' is changing slowly and that the curve f will have a small curvature. Note that $f'' = c$ is a parabola, while $curvature = c$ is a circle. So, $f'' \neq curvature$. But $f'' \approx curvature$, which has been pointed out by Strang's *Calculus*[25] that the most popular approximation of curvatures is the second derivative.

The second-order differential equation is hard to solve. But some special problems can be transformed into the first-order one.

Let us consider, for example, the initial value problem of the second-order linear differential equation:

$$\begin{cases} f'' + fa = 0, \\ f(0) = f_0, \quad f'(0) = v_0, \end{cases} \tag{3.1}$$

where the real number $a > 0$. Under the new parametric variable

$$\begin{cases} v = f', \\ v' = -fa, \end{cases}$$

Eq. (3.1) becomes a first-order matrix differential equation

$$\begin{cases} (f, v)' = (f, v) \begin{pmatrix} 0 & -a \\ 1 & 0 \end{pmatrix}, \\ (f, v)(0) = (f_0, v_0). \end{cases} \tag{3.2}$$

By imitating the scalar differential equation in Sec. 2.2.1 the matrix differential equation (3.2) will have the exponential solution

$$(f, v) = (f_0, v_0) e^{\begin{pmatrix} 0 & -a \\ 1 & 0 \end{pmatrix} x}, \tag{3.3}$$

where the right-hand side matrix is defined as the expansion

$$e^{\begin{pmatrix} 0 & -a \\ 1 & 0 \end{pmatrix} x} = \sum_{i=0}^{\infty} \frac{x^i}{i!} \begin{pmatrix} 0 & -a \\ 1 & 0 \end{pmatrix}^i. \tag{3.4}$$

This formula did not show us the oscillation (e.g., sines and cosines). How to display the oscillating solution? Recall the decomposition of the example e^{ix} in Section 1.9: Take use

$$i^2 = -1$$

then, splitting of e^{ix} into odd and even terms gives the cosines and sines:

$$e^{ix} = 1 + \frac{1}{1!}(ix) + \frac{1}{2!}(ix)^2 + \frac{1}{3!}(ix)^3 + \frac{1}{4!}(ix)^4 + \frac{1}{5!}(ix)^5 + \cdots$$
$$= \left(1 - \frac{1}{2!}x^2 + \frac{1}{4!}x^4 - \cdots \right) + 1 \left(\frac{1}{1!}x - \frac{1}{3!}x^3 + \frac{1}{5!}x^5 - \cdots \right)$$
$$= \cos x + i \sin x.$$

Now, we copy the above example: Make a diagonalization for the matrix:

$$\begin{pmatrix} 0 & -a \\ 1 & 0 \end{pmatrix}^2 = - \begin{pmatrix} a & 0 \\ 0 & a \end{pmatrix}.$$

$$
e^{\begin{bmatrix} 0 & -a \\ 1 & 0 \end{bmatrix} x} = \left(\begin{bmatrix} 1 & 0 \\ 0 & 1 \end{bmatrix} - \frac{x^2}{2!} \begin{bmatrix} a & 0 \\ 0 & a \end{bmatrix} + \cdots \right)
$$

$$
+ \begin{bmatrix} 0 & -a \\ 1 & 0 \end{bmatrix} \left(\frac{x}{1!} \begin{bmatrix} 1 & 0 \\ 0 & 1 \end{bmatrix} - \frac{x^3}{3!} \begin{bmatrix} a & 0 \\ 0 & a \end{bmatrix} + \cdots \right), \quad (3.5)
$$

in which the first term in the right-hand side displays cosines

$$
\begin{bmatrix} \cos(xa^{\frac{1}{2}}) & 0 \\ 0 & \cos(xa^{\frac{1}{2}}) \end{bmatrix}
$$

and the second term display sines

$$
\begin{bmatrix} 0 & -a \\ 1 & 0 \end{bmatrix} \left(\frac{x}{1!} \begin{bmatrix} 1 & 0 \\ 0 & 1 \end{bmatrix} - \frac{x^3}{3!} \begin{bmatrix} a & 0 \\ 0 & a \end{bmatrix} + \cdots \right)
$$

$$
= \begin{bmatrix} 0 & -a \\ 1 & 0 \end{bmatrix} \begin{bmatrix} a^{-\frac{1}{2}} & 0 \\ 0 & a^{-\frac{1}{2}} \end{bmatrix} \left(\frac{x}{1!} \begin{bmatrix} a^{\frac{1}{2}} & 0 \\ 0 & a^{\frac{1}{2}} \end{bmatrix} - \frac{x^3}{3!} \begin{bmatrix} a^{\frac{3}{2}} & 0 \\ 0 & a^{\frac{3}{2}} \end{bmatrix} + \cdots \right)
$$

$$
= \begin{bmatrix} 0 & -a^{\frac{1}{2}} \\ a^{-\frac{1}{2}} & 0 \end{bmatrix} \begin{bmatrix} \sin(xa^{\frac{1}{2}}) & 0 \\ 0 & \sin(xa^{\frac{1}{2}}) \end{bmatrix}
$$

$$
= \begin{bmatrix} 0 & -a^{\frac{1}{2}} \sin(xa^{\frac{1}{2}}) \\ a^{-\frac{1}{2}} \sin(xa^{\frac{1}{2}}) & 0 \end{bmatrix}.
$$

Finally, the exponential solution (3.4) displays the oscillating solution

$$
e^{\begin{pmatrix} 0 & -a \\ 1 & 0 \end{pmatrix} x} = \begin{pmatrix} \cos(xa^{\frac{1}{2}}) & -a^{\frac{1}{2}} \sin(xa^{\frac{1}{2}}) \\ a^{-\frac{1}{2}} \sin(xa^{\frac{1}{2}}) & \cos(xa^{\frac{1}{2}}) \end{pmatrix},
$$

and hence

$$
(f, v) = (f_0, v_0) \begin{pmatrix} \cos(xa^{\frac{1}{2}}) & -a^{\frac{1}{2}} \sin(xa^{\frac{1}{2}}) \\ a^{-\frac{1}{2}} \sin(xa^{\frac{1}{2}}) & \cos(xa^{\frac{1}{2}}) \end{pmatrix},
$$

i.e.,

$$\begin{cases} f = f_0 \cos(xa^{\frac{1}{2}}) + v_0 a^{-\frac{1}{2}} \sin(xa^{\frac{1}{2}}), \\ v = v_0 \cos(xa^{\frac{1}{2}}) - f_0 a^{\frac{1}{2}} \sin(xa^{\frac{1}{2}}). \end{cases}$$

J. Liu suggests to give a solver by guess:

$$(\sin xa^{\frac{1}{2}})'' = -a \sin xa^{\frac{1}{2}}$$
$$(\cos xa^{\frac{1}{2}})'' = -a \cos xa^{\frac{1}{2}}.$$

More interesting is the generalization of the above approach to a system of differential equations

$$\begin{cases} f'' + fA = 0, \\ f(0) = f_0, \quad f'(0) = v_0. \end{cases}$$

Its solver is just copied word for word from one equation example (3.1), only change is the positive number a to the real, positive, and symmetric matrix A. The solution is then

$$\begin{cases} f = f_0 \cos(xA^{\frac{1}{2}}) + v_0 A^{-\frac{1}{2}} \sin(xA^{\frac{1}{2}}), \\ v = v_0 \cos(xA^{\frac{1}{2}}) - f_0 A^{\frac{1}{2}} \sin(xA^{\frac{1}{2}}), \end{cases}$$

where $A^{\frac{1}{2}}$ is the square root of the positive matrix A and $A^{-\frac{1}{2}}$ is the inverse of the non-singular matrix $A^{\frac{1}{2}}$. See Hua's *Introduction of Higher Mathematics*.[11]

Remark. For a general second-order differential equation

$$af'' + bf' + cf = 0,$$

its solution should be of the type $e^{\lambda x}$ (see Sec. 1.8), where λ satisfies

$$a\lambda^2 + b\lambda + c = 0$$

with two roots: two real, two complex, or a real repeated, corresponding to the growing, decaying, or oscillating solutions. See Braun's *Differential Equations and Their Applications*[3] or Petrovski in Chap. 5 of Aleksandrov's *Mathematics*.[1]

3.2. Eigenvalue Problem

Consider the eigenvalue problem

$$-f''(x) = \lambda f(x), \quad f(0) = f(\pi) = 0,$$

which can be seen by a series of visible graphs

$$f_k(x) = \sqrt{\frac{2}{\pi}} \sin kx, \quad k = 1, 2, \ldots$$

(called normalized eigenfunctions) whose second derivatives (or roughly the curvatures) are the multiples of

$$\lambda_k = k^2, \quad k = 1, 2, \ldots$$

(called eigenvalues) of these graphs, respectively.

The explicit eigenvalues, k^2, are just used to examine the performance of the approximate method: Can we construct the lower approximation, upper approximation, and their expansions?

Let us first construct the lower approximation, e.g., the three-point finite difference method (of mesh size $2h$ rather than h), a simple approximate method, which has unchanged eigenfunctions

$$f_{k,h}(2ih) = \sin k(2ih), \quad 2h = \frac{\pi}{n}, \quad 1 \le i, \ k \le n - 1,$$

and the explicit eigenvalues

$$\lambda_{k,h} = \frac{2}{(2h)^2}(1 - \cos 2kh)$$

which are more complex than the continuous case, k^2, and different from the exact eigenvalues given below, just like the inscribed polygon approximation, π_n, different from π given below. Furthermore, we have an expansion: when $k \ll n$ and $h \ll 1$,

$$\lambda_{k,h} = \frac{2}{(2h)^2}(1 - \cos 2kh)$$

$$= \frac{2}{(2h)^2}\left(\frac{k^2(2h)^2}{2!} - \frac{k^4(2h)^4}{4!} + \frac{k^6(2h)^6}{6!} + \cdots\right)$$

$$= \lambda_k - \frac{k^4}{3}h^2 + \frac{2k^6}{45}h^4 + \cdots < \lambda_k.$$

We then obtain an extrapolation of high rate:

$$\frac{4\lambda_{k,h/2} - \lambda_{k,h}}{3} = \lambda_k - \frac{k^6}{90}h^4 + \cdots < \lambda_k.$$

The next step is the two-dimensional example

$$\begin{cases} -\Delta f = \lambda f & \text{in } \Omega, \\ f = 0 & \text{on } \partial\Omega \end{cases}$$

with rectangle domain $\Omega = (0, \pi) \times (0, \pi)$ and the boundary $\partial\Omega$. Similar to the one-dimensional model, we also have the explicit (normalized) eigenfunctions and eigenvalues:

$$f_{kl}(x, y) = \frac{2}{\pi} \sin kx \sin ly,$$

$$\lambda_{kl} = k^2 + l^2,$$

where λ_{kl}, however, may be of multiple eigenvalues (but λ_{kk}, $k \leq 4$, are still simple, and λ_{55} are triple). We can also construct the lower approximation, e.g., the five-point finite difference method (of mesh size $2h$ rather than h), a simple approximate method, which has unchanged eigenfunctions

$$f_{kl,h}(2ih, 2jh) = \frac{2}{\pi} \sin k(2ih) \sin l(2jh),$$

$$2h = \frac{\pi}{n}, \quad 1 \leq i, j, k, \ l \leq n - 1,$$

and the explicit eigenvalues

$$\lambda_{kl,h} = \frac{2}{(2h)^2}(2 - \cos 2kh - \cos 2lh)$$

$$= \lambda_{kl} - \frac{k^4 + l^4}{3}h^2 + \frac{2(k^6 + l^6)}{45}h^4 + \cdots < \lambda_{kl} \qquad (3.6)$$

which are more complex than the continuous case, $k^2 + l^2$, and different from the exact eigenvalues given below, just like the inscribed polygon approximation, π_n, different from π given below. From the above expansion, we obtain an extrapolation of high rate:

$$\frac{4\lambda_{kl,h/2} - \lambda_{kl,h}}{3} = \lambda_{kl} - \frac{k^6 + l^6}{90}h^4 + \cdots < \lambda_{kl}.$$

We emphasize that it is too accidental where the lower approximation has explicit expressions for eigenvalues.

For getting more lower and upper approximations for eigenvalue problems, we need a more general framework of the approximate method, containing nonconforming and conforming finite element methods, which needs, however, more preparation than required for the simple finite difference method. See Lin's *Finite Element Methods* for more details.[20]

3.3. Boundary Value Problem

We now turn to the curvature. We know that a curve at different points has different bent degrees, called "curvature." How to compute the curvature of a curve at a point? We use the second derivatives, f''. Note that $f'' = c$ is a parabola, while *curvature* $= c$ is a circle. So, $f'' \neq curvature$. But $f'' \approx curvature$, which has been pointed out by Strang's *Calculus*[25] that the most popular approximation of curvatures is the second derivative.

This section seeks the solution curve by its curvatures, read constrainedly as a bounded value problem

$$-f'' = g(x), \quad f(-a) = f(a) = 0, \tag{3.7}$$

where $g(x)$ is known, the height solution, $f(x)$, is to be found.

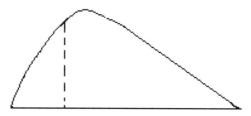

Recall that Euler used a continuous broken line to follow the solution of simplest differential equation (1.9). Can we use the broken line also to follow the solution of curvature equation (3.7)? Since the second derivatives of a broken line are piecewise zero, not the given f, we should avoid the second derivatives in Eq. (3.7) and use the weak equation.

3.4. Weak Equation

The weak equation is derived by means of test functions and integration by parts. Here, we just use a special test function: Let $\varphi_1(x)$ denote the following broken line

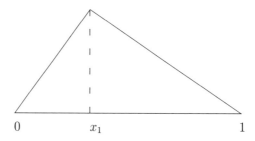

normalized with

$$\varphi_1(x_1) = 1$$

at a nodal point x_1, and satisfy the zero boundary condition

$$\varphi_1(0) = \varphi_1(1) = 0.$$

A general broken line can then be expressed by $c\varphi_1$, where c is the height of the broken line. Multiply both sides of (3.7) by the test function, e.g., φ_1, and integrate over $(0, 1)$:

$$-\int_0^1 f'' \varphi_1 dx = \int_0^1 g\varphi_1 dx, \tag{3.8}$$

and use integration by parts on the left-hand side to remove one derivative from f'' onto φ_1:

$$\int_0^1 f' \varphi_1' dx = \int_0^1 g\varphi_1 dx, \tag{3.9}$$

where we use the fact that φ_1 is zero at boundaries. Note (see Eriksson–Estep–Hansbo–Johson's *Computational Differential Equations*, pp. 116–117[9]) that the derivative φ_1' in (3.9) is a piecewise constant and is not defined at the node x_1. However, the integral with integrand $f'\varphi_1'$ is nevertheless uniquely defined as the sum of integrals over the subintervals $(0, x_1)$ and $(x_1, 1)$. This is due to the basic fact of integration that two functions that are equal except at a finite number of points have the same integral.

Nowadays we refer to (3.9) as a weak form of (3.7) when φ_1 goes through all test functions such that (3.9) makes sense. Equations (3.9) and (3.7) are then equivalent when f is smooth, but the weak one, (3.9), allows the broken line to be an approximation. This is the basis of the finite element method. In fact, the finite element method is the broken-line solver,

$$c_1 \varphi_1,$$

of weak equation (3.9).

An ideal broken-line solver occurs if $c_1 = f(x_1)$, i.e., the broken-line interpolation of the exact solution u:

$$f_I = f(x_1)\varphi_1, \tag{3.10}$$

which coincides at nodal point x_1 with f,

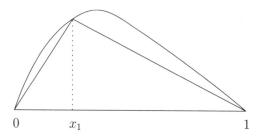

$$0 \qquad x_1 \qquad\qquad\qquad 1$$

but the solution value $f(x_1)$ is unknown. We can only use weak equation (3.9) for determining c_1.

3.5. Finite Element Solution and Interpolation

Through a technical computation, the finite element method is shown to be a secant line approximation of the solution curve.

We now substitute the broken-line solution $c_1\varphi_1$ into the weak equation (3.9) and obtain the constant c_1 satisfying

$$c_1 \int_0^1 \varphi_1'\varphi_1' = \int_0^1 g\varphi_1.$$

We hope c_1 is close to the exact solution $f(x_1)$. In fact, the right-hand side

$$\int_0^1 g\varphi_1 = \int_0^1 f'\varphi_1'$$
$$= f\varphi_1'|_0^{x_1} + f\varphi_1'|_{x_1}^1$$
$$= \int_0^1 f_I'\varphi_1'$$
$$= f(x_1) \int_0^1 \varphi_1'\varphi_1',$$

where we use integration by parts and the fact that

$$\varphi_1'' = 0 \quad \text{on } (0, x_1), (x_1, 1)$$

and the definition of broken-line interpolation f_I in (3.10). We then have

$$c_1 = f(x_1).$$

This is just the ideal case we expected in (3.10) where the broken-line solver coincides at the nodal point with the exact solution (a curve), i.e., the finite element method produces the exact solution at the nodal point without the discretization error. In other words, the finite element solution $f(x_1)\varphi_1$ equals the interpolation f_1 [see (3.10)] of the exact solution f.

Even this is an accidental phenomenon that can be used to understand the finite element method, like a secant approximation of the solution curve and, in the special case, the inscribed polygon approximation of a circle.

Since x_1 is an arbitrary point on interval $(0, 1)$, we have solved equation (3.10):

$$f(x_1) = c_1 = x_1(1 - x_1) \int_0^1 \varphi_1(x)g(x)dx,$$

where $x_1(1 - x_1)\varphi_1(x)$ is the so-called Green's function.

3.6. Generalization

The same idea can be generalized to the problem

$$-(pf')' = g, \quad f(0) = f(1) = 0 \tag{3.11}$$

with a piecewise constant coefficient

$$p = p_i \quad \text{on } (x_i, x_{i+1})$$

on the partition

$$0 = x_0 < x_1 < \cdots < x_{n-1} < x_n = 1.$$

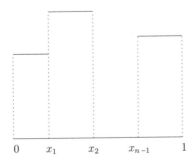

$$0 \quad x_1 \quad x_2 \quad x_{n-1} \quad 1$$

The simplest test function is now the broken line as follows:

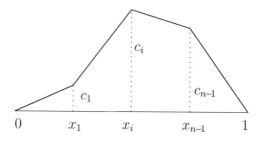

which can be described by nodal values c_i:

$$\varphi = (0, c_1, \ldots, c_i, \ldots, c_{n-1}, 0)$$

like a vector described by its components c_i ($i = 1, \ldots, n - 1$). Thus, a broken-line function can be imagined as a vector with "coordinate axes"

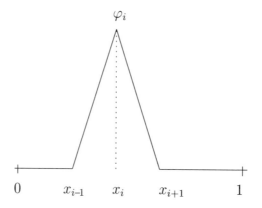

$$\varphi_1 = (0, 1, 0, \ldots, 0), \quad \varphi_i = (0, \ldots, 1, \ldots, 0), \quad \varphi_{n-1} = (0, \ldots, 0, 1, 0),$$

and the general broken line φ can then be written in the "coordinate" form

$$\sum_{i=1}^{n-1} c_i \varphi_i. \tag{3.12}$$

The weak form of (3.11) is written as

$$\int_0^1 p f' \varphi_j' = \int_0^1 g \varphi_j, \quad 1 \le j \le n, \tag{3.13}$$

and the finite element method is broken-line solver (3.12).

An ideal broken-line solver occurs if $c_i = f(x_i)$, i.e., the broken-line interpolation f_I of f:

$$f_I = \sum_{i=1}^{n-1} f(x_i)\varphi_i, \tag{3.14}$$

but the solution values $f(x_i)$ are unknown. We can only use weak equation (3.13): substitute broken-line solver (3.12) into (3.13) and obtain coefficients c_i satisfying the algebraic system

$$\sum_{i=1}^{n-1} c_i \int_0^1 p\varphi_i'\varphi_j'dx = \int_0^1 g\varphi_j dx, \quad 1 \le j \le n-1, \tag{3.15}$$

called the finite element equation. We hope c_i is close to the exact solution $f(x_i)$. In fact, the right-hand side

$$\int_0^1 g\varphi_j dx = \int_0^1 pf'\varphi_j'dx = \int_{x_{j-1}}^{x_{j+1}} pf'\varphi_j'dx$$
$$= p_{j-1}f\varphi_j'|_{x_{j-1}}^{x_j} + p_j f\varphi_j'|_{x_j}^{x_{j+1}}$$
$$= \int_0^1 pf_I'\varphi_j'dx = \sum_{i=1}^{n-1} f(x_i) \int_0^1 p\varphi_i'\varphi_j'dx,$$

where we use integration by parts and the fact that

$$\varphi_i'' = 0 \quad \text{on} \quad (0, x_{i-1}), (x_{i+1}, 1)$$

and the definition of f_I in (3.14). We then have

$$c_i = f(x_i), \quad 1 \le i \le n-1.$$

This is again an ideal but accidental phenomenon, where the finite element solution (a broken line) coincides at nodal points x_i $(0 \le i \le n)$ with the exact solution (a curve), i.e., the finite element method produces the exact solution at the nodal points without the discretization error. In other words, the finite element solution of Eq. (3.11),

$$\sum_{i=1}^{n-1} f(x_i)\varphi_i,$$

equals the interpolation f_I in (3.14) of the exact solution f.

Thus, at nodes x_i on $(0, 1)$, we have solved Eq. (3.11):

$$f(x_i) = c_i, \quad 1 \le i \le n - 1,$$

where $(c_1, \ldots, c_{n-1})$ is the solution of algebraic system (3.15).

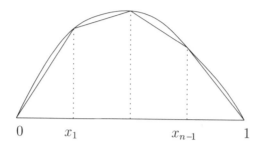

$$0 \qquad x_1 \qquad\qquad\qquad x_{n-1} \qquad 1$$

The above phenomenon is too good but accidental. If we go to problem (3.11) with a variable coefficient p (rather than the piecewise constants), or a differential equation of multivariables, the usual finite element solution no longer coincides with the exact solution at any interior point.[a] There is, actually, a difference between the finite element solution and the exact solution of general differential equations. Lin's *Finite Element Methods* discusses how to reduce the difference.[20]

3.7. Summary

Some characteristics of the finite element method have been exploited with the one variable differential equation, e.g.,

1. using broken-line or, more generally, piecewise polynomial functions;

2. using the weak equation.

The finite element method is the piecewise polynomial solution of the weak equation. In an ideal case, the finite element method is just an interpolation of the exact solution.

For more details, see, e.g., Refs. 4, 6, 13, 16–20.

---

[a] There is a paper by Hlavacek and Krizek discussing on exact results in some finite element methods.[10]

Chapter 4

Free Calculus

in Abstract Space

Outline: abstract calculus = free calculus with the replacement of
$| \cdot |$ *by* $\| \cdot \|$

4.1. Function Spaces, Norms, and Triangle Inequality

Functional analysis is too abstract for a beginner except that it can be imagined by plane geometry.

The object of differential equations is the functions, instead of a number or a vector. One may, however, imagine a function as a vector with infinitely many components — this is the basic idea of functional analysis. Once this is accepted, we may imagine the function space as plane geometry, discussing the triangle, its lengths and angles, triangle inequality, cosine inequality, etc. Let us detail this idea.

We have, when fixing a point as origin in the space, the notion of vectors, and the space with all vectors in it becomes a "vector" space. The study of this vector space is called geometry. Then why don't we consider a collection of functions and imagine it as a "function" space?

Let us recall that a vector in a two- or three-dimensional space is experssed as

$$f = (f_1, f_2, f_3),$$

with a length (or norm):

$$\|f\| = \sqrt{\sum_{i=1}^{3} f_i^2}.$$

Two vectors u and

$$g = (g_1, g_2, g_3)$$

form a triangle

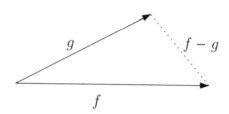

where the length of the third vector

$$f - g = (f_1 - g_1, f_2 - g_2, f_3 - g_3)$$

is shorter than the sum of the lengths of other two vectors (because the straight line is the shortest route between two points):

$$\|f - g\| \le \|f\| + \|g\|,$$

i.e., the triangle inequality written algebraically:

$$\sqrt{\sum_{i=1}^{3}(f_i - g_i)^2} \le \sqrt{\sum_{i=1}^{3} f_i^2} + \sqrt{\sum_{i=1}^{3} g_i^2}.$$

The algebraic language hints us further to imagine the multi-dimensional results. For example, consider

$$f = (f_1, f_2, \ldots, f_n), \quad g = (g_1, g_2, \ldots, g_n)$$

as the vectors in n-dimensional space and

$$\|f\| = \sqrt{\sum_{i=1}^{n} f_i^2}, \quad \|g\| = \sqrt{\sum_{i=1}^{n} g_i^2}$$

as their lengths. We may boldly imagine a triangle inequality

$$\sqrt{\sum_{i=1}^{n}(f_i - g_i)^2} \le \sqrt{\sum_{i=1}^{n} f_i^2} + \sqrt{\sum_{i=1}^{n} g_i^2},$$

although it needs an accurate proof.

The key idea in functional analysis is to boldly imagine a function out of many other functions as a vector out of many other vectors. This can easily be seen for a broken-line function f with n components,

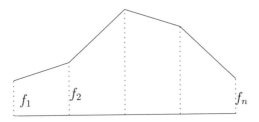

where

$$f = (f_1, f_2, \ldots, f_n). $$

As the number of components keeps on increasing, their limit will be a function f with infinitely many components,

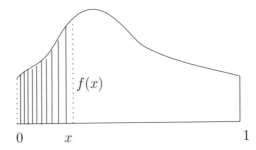

where

$$f = (f(x), \ 0 \le x \le 1).$$

Thus, function f is a vector with infinitely many components $f(x)$ when x runs over from 0 to 1.

Once function f is imagined as a vector with infinitely many components $f(x), 0 \le x \le 1$, its length should be

$$\|f\| = \sqrt{\int_0^1 f(x)^2 dx}.$$

The vector

$$f - g = (f(x) - g(x), \ 0 \le x \le 1)$$

taking the "distance" then satisfies the Minkowski inequality:

$$\sqrt{\int_0^1 (f(x) - g(x))^2 dx} \le \sqrt{\int_0^1 f(x)^2 dx} + \sqrt{\int_0^1 g(x)^2 dx}, \qquad (4.1)$$

a desired property for a geometry of function space although it needs an accurate proof. The most important thing is that we know how to envision the Minkowski inequality.

Could you believe that the function space notion could turn the Minkowski inequality into a triangle inequality (i.e., the straight line is the shortest route between two points)? Boldly? Is this playing a magic?

In summary, we can introduce an abstract function space, the linear norm space, as follows:

let $f, \ g, \dots$ be the elements of a real linear space with the norm $\| \cdot \|$, a number, satisfying

$$\|f\| \ge 0; \quad \|f\| = 0 \Leftrightarrow f = 0;$$
$$\|\alpha f\| = |\alpha| \|f\| \quad \forall \alpha, \text{ a real number};$$
$$\|f + g\| \le \|f\| + \|g\|.$$

4.2. Angle and Schwartz's Inequality

Let two vectors f and g form an angle θ:

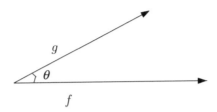

The cosine theorem

$$\|f - g\|^2 = \|f\|^2 + \|g\|^2 - 2\|f\|\|g\| \cos \theta$$

leads to

$$\cos \theta = \frac{\|f\|^2 + \|g\|^2 - \|f - g\|^2}{2\|f\|\|g\|}$$

or, in an algebraic language,

$$\cos \theta = \frac{\sum f_i g_i}{\sqrt{\sum f_i^2}\sqrt{\sum g_i^2}},$$

or

$$\cos \theta = \frac{\displaystyle\int_0^1 f(x)g(x)dx}{\sqrt{\displaystyle\int_0^1 f(x)^2 dx}\sqrt{\displaystyle\int_0^1 g(x)^2 dx}}.$$

If we still accept the cosine inequality

$$|\cos \theta| \le 1$$

in the multi-dimensional space or the function space, we should accept the Schwartz inequality:

$$\left|\sum f_i g_i\right| \le \sqrt{\sum f_i^2}\sqrt{\sum g_i^2},$$

or

$$\left| \int_0^1 f(x)g(x)dx \right| \le \sqrt{\int_0^1 f(x)^2 dx} \sqrt{\int_0^1 g(x)^2 dx} \qquad (4.2)$$

although they need accurate proofs. The most important thing is that we know how to envision them before proving.

Could you believe that the function space notion could turn the Schwartz inequality into $|\cos\theta| \le 1$? Boldly? Is this playing a magic?

4.3. Inner Product

To get an accurate proof for (4.2), let us introduce a convenient notation

$$(f, g) = \sum f_i g_i \overset{\text{or}}{=} \int_0^1 f(x)g(x)dx, \qquad (4.3)$$

called a (real) inner product; it is similar to the real number product:

$$(f, f) \ge 0, (f, f) = 0 \Leftrightarrow f = 0; \qquad (4.4)$$
$$(f, g) = (g, f); \qquad (4.5)$$
$$(f + w, g) = (f, g) + (w, g); \qquad (4.6)$$
$$(\alpha f, g) = \alpha(f, g) \qquad \forall \alpha, \text{ a real number.} \qquad (4.7)$$

These properties guarantee conclusion (4.2): Define a norm $\|\cdot\|$ induced by the inner product

$$(f, f) = \|f\|^2. \qquad (4.8)$$

Then, for unit vectors f and g,

$$\|f\| = \|g\| = 1,$$

we have by (4.8), (4.6), and (4.5)

$$\|f \pm g\|^2 = \|f\|^2 \pm 2(f, g) + \|g\|^2$$
$$= 2[1 \pm (f, g)] \ge 0, \qquad (4.9)$$

and hence

$$|(f, g)| \le 1.$$

For general vectors f and g, we have by (4.7)

$$\left| \left(\frac{f}{\|f\|}, \frac{f}{\|g\|} \right) \right| \leq 1$$

or

$$|(f, g)| \leq \|f\| \|g\| \qquad (4.10)$$

which proves the Schwartz inequality (4.2). Thus, the algebraic language continues the college analysis to a few lines!

The Schwartz inequality (4.10) leads to the triangle inequality (4.1):

$$\begin{aligned}
\|f - g\|^2 &= \|f\|^2 - 2(f, g) + \|g\|^2 \\
&\leq \|f\|^2 + 2\|f\| \|g\| + \|g\|^2 \\
&= (\|f\| + \|g\|)^2.
\end{aligned}$$

4.4. Orthogonality and Projection

Having the angle notion we may define the orthogonality for two vectors f and g:

$$(f, g) = 0 \iff \cos \theta = 0,$$

and an algebraic formula, (4.9), rigorously proves a big theorem, the Pythagoras theorem:

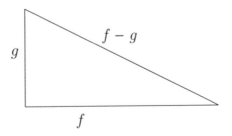

$$\|f - g\|^2 = \|f\|^2 + \|g\|^2.$$

Without the orthogonal condition we may have the parallelogram formula:

$$\|f - g\|^2 = 2\|f\|^2 + 2\|g\|^2 - \|f + g\|^2,$$

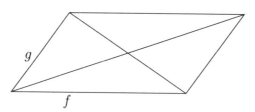

which is crucial for some convergence arguments (Cauchy sequence).

We may introduce the notion of projection. If a vector f has an orthogonal projection φ onto the subspace $V : \varphi \in V$ such that

$$(f - \varphi, g) = 0 \quad \forall g \in V,$$

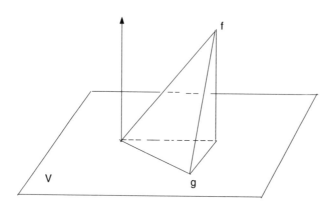

then such a φ should be the vector in V closest to f:

$$\|\varphi - f\| = \min_{g \in V} \|g - f\|.$$

The proof is based on the orthogonality:

$$(f - \varphi, g - \varphi) = 0,$$

so

$$\|f - g\|^2 = \|f - \varphi\|^2 + \|g - \varphi\|^2 \geq \|f - \varphi\|^2 \quad \forall g \in V.$$

The above orthogonality and minimum properties from a geometric imagination are basic for the finite element method described in Chap. 3.

The orthogonality and projection are also the basic ideas for the Fourier integral representation: if $\varphi_0, \varphi_1, \varphi_2, \varphi_3, \varphi_4, \ldots$ are an orthogonal, normalized, and complete (formally, for the moment, since the notion of completeness has not been made precise) coordinates axes (e.g., $\varphi_0 = \frac{1}{\sqrt{2\pi}}, \varphi_1 = \cos t, \varphi_2 = \sin t, \varphi_3 = \cos 2t, \varphi_4 = \sin 2t, \ldots$), then

$$f = \sum_i (f, \varphi_i)\varphi_i.$$

4.5. Different Inner Products and Norms

Inner product, besides (4.3), may take different definitions, e.g.,

$$(f, g)_k = \int_0^1 \sum_{\alpha=0}^k D^\alpha f D^\alpha g \, dx,$$

called H^k-inner product, or L^2-inner product in case $k = 0$; the induced norm,

$$\|f\|_k = \sqrt{\int_0^1 \sum_{\alpha=0}^k |D^\alpha f|^2 dx}$$

called H^k-norm, or L^2-norm in case $k = 0$.

Summary. Based on the imagination that a function is a vector with infinitely many components, some results in analysis are paralleled to plane geometry, cf. Gelfand in Chap. 18 of Aleksandrov's Mathematics.

So far we have introduced only the formal notions in functional analysis—just enough to establish the fundamental formula in the next section. For precise notions and complete theory, such as the weak derivative in Sobolev space H^k and "completeness" of a function space, etc., readers are referred to Brenner and Scott.[4]

4.6. Abstract Calculus

Consider the abstract function $f(x)$ in a general linear normed space (without completeness, just a linear algebra) with the independent variable x still being the real number (e.g., a vector function of the real variable). We recall that the original proof of the abstract FT in Ljusternik–Sobolev[22] (Chapter 8) is based on the Hahn–Banach theorem and completeness, and uses priori knowledge from real-valued calculus; their proof is very clever and long. However, they are avoided in this section. The abstract FT here can be rigorously proved through a definition itself and a few lines of arithmetic: Fundamental essences of

$$\text{abstract calculus} = \text{a definition} + \text{few lines of arithmetic}$$

without *a priori* knowledge and tricky techniques. Where, of course the absolute value $|\cdot|$ must be replaced by the norm $||\cdot||$. See Ref. 19.

We shall not repeat the material in Chapter 1 here but just mention the case when the function f is smooth.

Then, the first (differential) inequality becomes. With node x and its nearby points $x + h$

$$\|f(x+h) - f(x) - f'(x)h\| \le \epsilon(h)h \le Ch^2$$

and the second (fundamental) inequality becomes

$$\|f(b) - f(a) - \text{sum of} f'(x)h\| \le (b-a)\epsilon(h) \le Ch$$

without the ϵ language and with better accuracies.

Appendix

Professor Qun Lin, one of the honorary editors of our journal, is a researcher in numerical partial differential equations. He is also a mathematics educator in teaching. He has supervised a number of excellent students who are now active researchers around the world, from pure mathematics to applied mathematics and scientific computing. A special issue of *International Journal of Information & Systems Sciences, Volume 2, Number 3, 2006*, is dedicated to his 70th birthday.

Professor Lin has also made a contribution to the universal education. For example, he spreads an elementary knowledge of functional analysis in many colleges in China since 1990's. He gave an elementarization of functional analysis calculus. His idea is to use two elementary inequalities to define the derivative and represent the fundamental theorem, respectively. To be more precisely let f be an abstract function defined on $[a, b]$ and taken values in a linear norm space (without the completeness notion, just an elementary linear algebra). The derivative f' is defined with an elementary inequality: For all $x + h$ near x,

$$||f(x + h) - f(x) - f'(x)h|| \leq \epsilon(h)h$$

where the notation $\epsilon(h)$ depends only on the size of the variable increment h and is chosen small so that the derivative f' is uniquely defined. Then adding up these inequalities on each subinterval $[x, x + h]$ gives another elementary inequality: The fundamental theorem,

$$||f(b) - f(a) - \sum_{x \in nodes} f'(x)h|| \leq (b - a)\epsilon(h)$$

which can be used to define the definite integral. Such an elementary definition of the derivative, and such an elementary presentation and proof of the fundamental theorem, are much easier for the student to understand.

On this occasion, we wish Professor Lin of safety, joviality, and longevity.

Zhangxin Chen
Southern Methodist University
Dallas, Texas

CALCULUS OF FUNCTIONAL ANALYSIS BECOMES ELEMENTARY ALGEBRA

QUN LIN

Abstract. Calculus of functional analysis becomes two elementary inequalities, first for the derivative definition and second for the fundamental theorem, without using the limit notion, completeness and the Hahn-Banach theorem.

Key Words. Functional analysis, calculus, elementary inequality.

1. Introduction

Calculus of functional analysis, including the derivative definition and the fundamental theorem (FT), uses the $\epsilon - \delta$ notion, completeness and the Hahn-Banach theorem. They are to be avoided in this note: Derivative is defined by an elementary inequality with an error bound (see (1) below) and without using $\epsilon - \delta$ at the beginning, and the FT is then given by a second inequality (see (2) below) which is nothing but the sum of former inequalities themselves (see the proof of (2) below) without using completeness and the Hahn-Banach theorem. Calculus of functional analysis is indeed reduced into an inequality.

2. Derivative definition becomes an elementary inequality

Let f be an abstract function (producing an abstract curve) defined on an interval $[a, b]$, containing subinterval $[x, x + h]$ and taken values in a linear norm space (without the completeness notion, just an elementary linear algebra). The height variation over $[x, x + h]$, $f(x + h) - f(x)$, is computed by differential = derivative $\times$ base $= f'(x)h$, with an error bound $\epsilon(h)h$: For all $x + h$ near x,

$$(1) \qquad ||f(x + h) - f(x) - f'(x)h|| \leq \epsilon(h)h$$

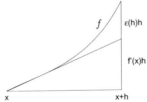

the height $f(x + h) - f(x)$ over subinterval $[x, x + h]$ is computed by its linear part, differential $f'(x)h$, with error $\epsilon(h)h$

2000 *Mathematics Subject Classification.* 00A, 65D and 97U.

where the notation $\epsilon(h)$ depends on the size of the argument increment h but is independent of the argument x, and is chosen small so that the derivative f' is uniquely defined.

3. Fundamental theorem becomes another elementary inequality

If there exists a function f' and a bound $\epsilon(h)$ satisfying the first inequality, (1):

$$||f(x+h) - f(x) - f'(x)h|| \leq \epsilon(h)h$$

then, adding up these inequalities on each subinterval $[x, x+h]$

gives a second elementary inequality: The total height over $[a, b]$ is computed by Reimann's sum with an error proportional to $\epsilon(h)$ in the first inequality:

$$(2) \qquad ||[f(b) - f(a)] - \sum_{x \in nodes} f'(x)h|| \leq (b-a)\epsilon(h)$$

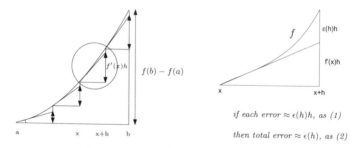

if each error $\approx \epsilon(h)h$, *as (1)*

then total error $\approx \epsilon(h)$, *as (2)*

this is why we define the differential as (1)

or even

$$(3) \qquad ||[f(b) - f(a)] - \sum_{\xi \in [x, x+h]} f'(\xi)h|| \leq (b-a)\epsilon(h).$$

When $\epsilon(h)$ is chosen small then the total error, (3), is still small, so the element $f(b) - f(a)$ depends on the end points a and b and is independent of the division for the subintervals, and can then be used to define the definite integral $\int_a^b f'(x)dx$. This is the FT. For a detailed proof of (2), see the following familiar argument in real calculus (e.g. [1–4]).

Let the total interval $[a, b]$ be divided into $n + 1$-equal subintervals $[x_i, x_i + h]$, where $a = x_0 < x_1 < \cdots < x_{n+1} = b$, $h = x_{i+1} - x_i$.

Use the uniform inequality (1) to each subinterval:

$$f(x_0 + h) - f(x_0) - f'(x_0)h = h\epsilon(h, x_0)$$
$$f(x_1 + h) - f(x_1) - f'(x_1)h = h\epsilon(h, x_1)$$
$$\cdots = \cdots$$
$$f(x_n + h) - f(x_n) - f'(x_n)h = h\epsilon(h, x_n)$$

satisfying

$$\text{upper}_x \|\epsilon(h, x)\| \leq \epsilon(h),$$

and add up. The left sides add to

$$f(b) - f(a) - \sum_{i=0}^{n} f'(x_i)h$$

and the right sides add to the total error

$$\|\sum_{i=0}^{n} h\epsilon(h, x_i)\| \leq (b - a)\epsilon(h)$$

or

$$\|[f(b) - f(a)] - \sum_{i=0}^{n} f'(x_i)h\| \leq (b - a)\epsilon(h).$$

This completes the proof of (2). By the same argument, (3) can be proved with more calculation. Notice that the above argument uses only the definition (1) itself without using completeness and the Hahn-Banach theorem.

Notice also that the uniform definition (1) (e.g. [1–4]):

$$f(x + h) - f(x) - f'(x)h = \epsilon(h, x)h$$

$$\text{upper}_x \|\epsilon(h, x)\| \leq \epsilon(h)$$

is a small modification from the pointwise definition (see Ljusternik-Sobolev, Chapter 8)

$$f(x + h) - f(x) - f'(x)h = \epsilon(h, x)h$$

$$\text{at fixed } x : \|\epsilon(h, x)\| \leq \epsilon(h)$$

but it greatly simplifies regular calculus of functional analysis.

Functional analysis teachers might react the uniform condition (1). Indeed, it has already been proved by Vainberg that the uniform condition (1) is equivalent to uniform continuity of the pointwise derivative. Such a uniform continuity is exactly the standard condition in the FT, but using the uniform condition (1) is much more simple and practical, reducing functional analysis into elementary algebra.

This is an elementary approach and radically simplifies the proof of the FT for functional analysis, where functional analysts (see Ljusternik-Sobolev, Chapter 8) persistently used Hahn-Banach's theorem and so it is tricky, complicated and even not complete. We use such an approach to spread functional analysis since 1990's.

Acknowledgments

Thanks to Professors Zhangxin Chen, Dan Velleman, Congxin Wu and Hung Hsi Wu for their criticisms. Professor Jingzhong Zhang has made full use of the uniform inequality (1) itself to complete an elementarization of differential calculus very successfully.

References

[1] Lin, Q., Calculus Cartoon, Guangming Daily, China, June 27, 1997, and Renming Daily, China, August 6, 1997.

[2] Lin, Q., Calculus Cartoon, Guangxi Normal University Press, Guilin, Jan. 1999.

[3] Calculus for Humanities, Lin, Q. (eds.) and Wu, C., Hebei University Press, 2002.

[4] Lin, Q., Differential Equations and the Trigonometry Measurement, Tsinghua University Press, 2005.

[5] Ljusternik, L. and Sobolev, V., Elements of Functional Analysis, 1965.

[6] Vainberg, M., A Hamerstein theorem for nonlinear integral equations, Moscow University Transaction, Vol.100:1, 93-103, 1946.

[7] Vainberg, M., Variational Methods for the Study of Nonlinear Operators, Moscow, 1956.

LSEC, ICMSEC, Academy of Mathematics and Systems Science, Chinese Academy of Sciences,Beijing 100080, China

E-mail: linq@lsec.cc.ac.cn

Bibliography

1. A. Aleksandrov, *Mathematics, Its Essence, Methods and Role* (Publishers of the USSR Academy of Sciences, Moscow, 1956).
2. V. Arnold, Why should we study mathematics? Quanta 5–15 (1993).
3. M. Braun, *Differential Equations and Their Applications* (1978).
4. S. Brenner and R. Scott, *The Mathematical Theory of Finite Element Methods* (Springer-Verlag, New York, 1994).
5. C. Bruter, *Sur La Nature Des Mathematiques* (Gauthier-Villars, Paris, 1973).
6. Z. Chen, *Finite Element Methods and Their Applications* (Springer, 2005).
7. R. Courant and F. John, *Introduction to Calculus and Analysis* (Springer, 1989).
8. K. Dovermann, *Applied Calculus*, July 1999 [http://www.math.hawaii.edu/%7Eheiner/calculus.pdf].
9. K. Eriksson, D. Estep, P. Hansbo and C. Johson, *Computational Differential Equations* (Cambridge University Press, 1996).
10. I. Hlavacek and M. Krizek, On exact results in the finite element method, *Appl. Math.* **46**, 467–478 (2001).
11. L. Hua, *Introduction of Higher Mathematics* (Science Press, 1979).
12. H. Karcher, *Analysis mit gleichmgen Fehlerschranken*, Okt. 2002 [http://www.math.uni-bonn.de/people/karcher].
13. M. Krizek and P. Neittaanmaki, *Finite Element Approximation of Variational Problems and Applications*, Pitman Monographs and Surveys in Pure and Applied Mathematics, Vol. 50, 1990.

14. P. Lax, S. Burstein and A. Lax, *Calculus with Applications and Computing* (Springer-Verlag, New York, 1976).

15. Q. Lin, Calculus cartoon, *Guangming Daily*, China, June 27, 1997, and *Renming Daily*, China, August 6, 1997.

16. Q. Lin, *Calculus Cartoon* (Guangxi Normal University Press, Guilin, 1999).

17. Q. Lin, From trigonometry to calculus, English print, 2001.

18. Q. Lin and C. Wu (eds.) *Calculus for Humanities* (Hebei University Press, 2002).

19. Q. Lin, Calculus of functional analysis becomes elementry algebra, International Journal of Information and System Sciences, Vol. 2(3), 281–284, 2006.

20. Q. Lin and J. Lin, *Finite Element Methods: Accuracy and Improvement* (Science Press, 2006).

21. M. Livshits, *Simplifying Calculus by Using Uniform Estimates*, 2004 [mysite.verizon.net/michaelliv www.mathfoolery.org].

22. L. Ljusternik and V. Sobolev, *Elements of Functional Analysis*, 1965.

23. M. Ryan, *Calculus for Dummies* (Wiley Publishing, Inc., 2003).

24. J. Stewart, *Calculus*, 5th Edn. (Brooks/Cole, CA, 2003).

25. G. Strang, *Calculus* (Wellesley-Cambridge Press, Wellesley, MA, 1991).

26. L. Tolstoy, *War and Peace*, Translated by Louise and Aylmer Maude, 1994 [contains the comments about calculus method in pp. 879–880].

27. M. Vainberg, A Hamerstein theorem for nonlinear integral equations, *Moscow Univ. Trans.* **100**(1), 93–103 (1946).

28. M. Vainberg, *Variational Methods for the Study of Nonlinear Operators*, Moscow, 1956.

29. L. Wen, *One Variable Calculus* (Shanghai Science and Technology Press, Shanghai, 1981).

30. J. Zhang, *Elementarization of Calculus*, 2006.

31. X. Zhu and J. Zheng, *Calculus I, II*, eds. S. Xiao, Y. Wang, *et al.* (Higher Education Press, 2000).